ON COMPARING AND EVALUATING SCIENTIFIC THEORIES

POZNAŃ STUDIES
IN THE PHILOSOPHY OF THE SCIENCES AND THE HUMANITIES

VOLUME 72

This book series is partly sponsored by Adam Mickiewicz University, Poznań

The address: **prof. L. Nowak, Cybulskiego 13, 60-247 Poznań, Poland.**
Fax: (061) 8477-079 or (061) 8471-555
E-mail: epistemo@main.amu.edu.pl

ON COMPARING AND EVALUATING SCIENTIFIC THEORIES

Edited by

Adam Jonkisz
and Leon Koj

Rodopi

Amsterdam – Atlanta, GA 2000

The paper on which this book is printed meets the requirements of "ISO 9706:1994, Information and documentation - Paper for documents - Requirements for permanence".

ISSN 0303-8157
ISBN: 90-420-1263-3 (bound)

Printed in The Netherlands

CONTENTS

PREFACE

In every-day activities in science and the humanities we very often compare and evaluate hypotheses, theories, and whole disciplines. When people say that, for example, the humanities represent a much lower stage of development than science, or that science contributes almost nothing to solving existential human problems while the humanities are very helpful in this field, they are comparing whole sets of sciences and theories. Comparisons of this kind are made mostly in informal conversation. In print, they are common in reviews.

The ways and criteria of differentiating between theories have not been systematically scrutinized. One can evaluate and compare theories with respect to hundreds of attributes (more general, more precise, explanatorily useful, influential, and so forth). There have been some very interesting attempts to put forward a theory of theory comparison but as a rule they are confined to theories in science and have little in common with comparing of conceptions in the humanities. Even in the case of theories in sciences not all comparisons are taken into account. The interest of methodologists is mostly directed at comparing successive theories in the same domain of research. In this restricted field some theories of comparison have been invented (K. Popper, I. Lakatos, L. Laudan, I. Niiniluoto and the structuralists may be mentioned here).

We would like the present book to be a kind of contribution to a more general theory of theory comparison, i.e. a theory that goes beyond the realm of mathematics and science. The authors of the essays collected in this volume describe – from a systematic as well as a historical point of view – various conditions, difficulties and ways of comparing and evaluating theories, using different metascientific concepts and views (e.g. the concept of system in the Bertalanffyan sense, the concept of explanatory power, the model of multidimensional social choice, some AI results, and the structuralist view). Hence, the title *On Comparing and Evaluating Scientific Theories* reflects the general idea of this collection.

Cieszyn and Lublin, June 1999 *Adam Jonkisz, Leon Koj*

Poznań Studies in the Philosophy of the Sciences and the Humanities
2000, Vol. 72, pp. 9-35

Leon Koj

For my Mother

METHODOLOGY AND VALUES[1]

ABSTRACT. In the present paper the methods of justification are conceived of as functions of factors of the following kind: sets of earlier accepted sentences, sets of consequence relations, sets of research instruments (books and papers included), sets of activities that render the acceptance of sentence as a result, etc. Different kinds of justification exhibit different numbers of factors that assume different degrees of importance. It is clear that there is an infinity of methods of justification. The choice between them is called a scientific strategy. This strategy is also a function from factors such as utility, truth, uniform world view, conformance with the traditional habits of a society, etc. These factors are of different importance in different strategies. It will be shown that methodology concerns methods of justification in scientific strategies and that necessarily methodology takes into account the importance and the different values of the above factors. Methodology is not value-free and the comparison of values and of the factors is the basis for all the ways of comparing the sciences and the humanities.

Introduction

What can be the purpose of investigating the relations between methodology and values? Does such a project not amount to comparing objects belonging to domains so different that they are almost incompatible? Contrary to what may appear at first sight, an investigation of the connections between methodology and values can turn out to be important. We shall consider two of the many topics regarding this problem: (1) Does methodology investigate values, that is, are values objects of investigation in methodology? (2) Does methodological investigation make use of values as a means of evaluating scientific research? The answers to these questions are diverse, highly controversial, and have broad implications concerning the importance of science and its results.

---

[1] The present paper was originally written as a part of research project RPBP III-22-4.

Before we investigate the above problems further, it is essential to realize the importance of solutions concerning values in sciences and methodology.

As is generally known, science is often treated as a domain of our activities whose purpose is to give us true information, objective information or, at least, intersubjective information. In other words, science is conceived as a body of such information. Methods used in the sciences should not, therefore, involve anything subjective, unverifiable, or arbitrary in other ways. Methodology is to be a science as well, so it ought not to incorporate any subjective components either. Values and evaluations resulting from them are, in some views, subjective and boil down to more or less overt manifestations of feelings and personal attitudes. Consequently, first-order sciences and methodology – as a metascience – ought not to make use of evaluations and values in their methods (that is, of subjective and arbitrary components). First order sciences and methodology are not allowed – in this view – to provide evaluations. They can only give brief and general descriptions of phenomena under investigation. Such a simplified picture of science was advanced by Positivism.

There are, however, other views on science and methodology. Some methodologists aim to describe science as it is, without advancing postulations and giving rules determining what it ought to be like. They notice, though, that we do come across values and evaluations in sciences, sometimes quite often. Scientists talk about important objects of study, insignificant deviation from predictions, etc. Many such types of statements can be found especially in the humanities[2]. Methodology is also full of evaluations of theories, methods, and of actual cognitive activities. If we correlate such a state of affairs with the view that evaluations are subjective and are expressions of personal attitudes and feelings, then science, and especially methodology, cannot be treated as domains of intersubjective or objective activities or results. In this case, we can only treat them as subjective, relative and dependent on factors of cognitive, cultural, and volitional nature, and also on interests, superstitions, etc.

The two pictures of science and methodology we presented above recognize the role of evaluations and values but they assign them a rather pitiful role: that of demolishing the objectivity of science and methodology. On the other hand, values and evaluations resulting from them liberate us from viewing the world in a purely rational manner, which deprives it of personal elements by eliminating evaluative and

---

2 In the present paper by "science" we mean both "hard" sciences and humanities.

emotional attitudes. We would otherwise face a lifeless and insensible world in which life would not be worth living.

We may conclude that relating methodology to values is not an arbitrary move but that it is one of the keys to understanding contemporary methodological and metamethodological discussions. The subject of these discussions is the objectivity or intersubjectivity of science or, alternatively, its subjectivization, relativization, or even irrationalization.

In investigating the connections between methodology and values we shall pay little attention to values – to what they are, how they can be recognized, etc. We shall be concerned mostly with methodology. As far as values are concerned, we shall only assume the following: objects, phenomena, events, and properties of objects that we evaluate are values in the respect under which they are being evaluated, or they possess value in this respect. Objects, phenomena, etc. which are values or have values, influence choices or actions. We further assume a knowledge of evaluative vocabulary. All other problems concerning values, like their classification, shall be taken from general axiology, should the need arise.

An introductory account of methodology

To investigate the relations between methodology and values, we shall agree on what we are going to mean by methodology in the present article. It seems there is no definition of methodology that everybody can agree upon. Therefore let us agree, without getting involved in countless discussions, on our understanding of this concept. We shall offer an explanation justifying our agreement.

It is generally believed that methodology is the realm of thought concerning science (not everybody agrees, though, to call methodology a science of sciences). Apart from methodology, there are other disciplines that investigate science; like history, sociology, psychology, and the organization of science. There also exist related branches inquiring into science, at least to some extent, like praxeology, theory of systems, theory of probability, logic, decision theory, etc. An account of methodology should enable us to distinguish it from these disciplines. It is also postulated that an account of methodology should embrace all disciplines, including those whose status is sometimes questioned, for example various types of humanistic investigations.

Our agreement on the meaning of the term "methodology" involves two things. Firstly, it involves the subject of methodological research.

Secondly, it concentrates on the problem of the language of methodology, and this point is much more extensive and rather more important.

Methodology differs from other types of inquiry into science, such as history, psychology, and sociology of science, mainly in that it deals with the justification of information acquired in science. To this end – because different components making up science contribute in different ways to the ultimate justification of the information – methodology examines the components of which science is made: for example basic propositions, definitions of concrete, dispositional, and theoretical concepts, descriptive laws, theories and sets of theories constituting science. After listing and describing these components, methodology examines different means of justification and the influence of these components on the justification. In this respect, logical relations among the components of science are taken into account (they are dealt with by the part of methodology that I would like to call the logic of science). Research procedures and methods of conducting them are also taken into account. These procedures go beyond what can be described by a logician. These are procedures like observation, conducting experiments, applying measuring devices, etc. This domain can be called, using a term introduced by Ajdukiewicz, methodological pragmatics. Contemporary science had its own earlier stages, which sometimes differed radically from its present state. The problem of the influence of earlier stages of science on its later forms belongs to methodological dynamics. The three types of methodology investigate their subjects of research consistently from the point of view of their influence on justification of scientific information.

In our brief description of the problems of methodology we have not attempted to say just what justification is. We have used the term "justification" in its naive, protoscientific, ambiguous meaning. The reason for this ambiguity is the contested status of the concept of justification. During further considerations there shall be some additional determinations of the term "justification." For similar reasons we have not tried to give an account of science. The picture of science shall emerge gradually, from various remarks concerning the methods of justification.

The language of methodology

We shall base our account of the language of methodology on its interpretation advanced by A. Tarski. According to Tarski, primitive terms of the language of methodology are divided into four groups:

1) names of simple statements of a discipline under investigation, 2) translations of expressions belonging to the language of a discipline under investigation, 3) specific terms of methodology (in Tarski it was only the sign of concatenation), 4) nonspecific terms, that is terms of formal logic. Tarski's remarks concern only the language of methodology of deductive sciences. We shall extend his scheme to the degree required to make it possible to characterize other sciences as well. The four categories must belong to the language of methodology for simple reasons. If we want to talk about values of some scientific statements, about their truth, probability, or verisimilitude, then we must say something about the statements themselves; this is why we need their names. We also need to say something about reality; this is why we need translations of scientific statements. Original scientific statements do not enter methodology for fear of antinomies. Additionally, we have the terms of logic – if they are not already incorporated in the first-order language of the discipline under investigation – and have specific expressions regarding that, which sciences themselves do not talk about, i.e. cognitive procedures, procedures of building object language. Using his terminology, Tarski defines several important concepts in methodology: the concept of truth, designation, consequence, axiom, system, consistency, model, and many others. But not all concepts of interest to us can be found in Tarski, like those of probability, verisimilitude, or corroboration. The language he adopted is too limited to give an account of them and this is why Tarski's language scheme shall be elaborated.

1. Let us consider the first element of the language of methodology, names of expressions belonging to the language of a discipline under investigation. Names of expressions are introduced to the language of methodology because expressions belonging to science are the bearers of information acquired in the course of scientific investigation. The expressions themselves are somewhat less interesting because the same information can be expressed in different kinds of languages, both natural and symbolic. However, because the expressions are bearers of information, it is necessary to talk about them. And it was generally believed that all sound knowledge could be expressed by a language. This position was stressed by Twardowski and Ajdukiewicz. The assumption that sound scientific knowledge is always expressed by a language has been questioned in methodology since the times of M. Polanyi. Originally, poets and mystics opposed this assumption. Currently, the talk about tacit knowledge refers to L. Fleck. Within the framework of Tarski's methodology there is no place for such knowledge because in the language of his methodology there is no place for information not expressed in a language. If we do not want to reject all considerations of extralinguistic (tacit) knowledge just by adopting a particular language of methodology, then we have to add to its terminological inventory a category of names for unvoiced information, besides names of expressions. We do not want to prejudice anything in advance, and therefore we need to add the names of pieces of unvoiced information (knowledge) besides the names of expressions. We do not have to assume, however, that they are not empty names, for we do not insist on a positive conclusion in this respect. This new, tentatively postulated kind of name creates some problems, for how can one construct the names of pieces of unvoiced and therefore unsegmented information? What name should be assigned to a piece of information contained by an unvoiced feeling? How can we do this in a way enabling one to distinguish one such piece of information from another, thereby assigning it an unequivocal name? Maybe our actions and abilities are linked to unvoiced information and maybe by their means it will be possible to assign names to such pieces of information.

When Tarski presented his understanding of methodology, what he had in mind were names of expressions built strictly according to rules of logical syntax, both to actual rules and to all the rules it was possible to construct. Actual scientific statements usually diverge from the forms prescribed by rules – especially by rules of natural language. For this reason, statements of methodology, conceived in Tarski's way, are idealizations and they go beyond actually formed expressions. Idealization necessarily involves – as is generally known – simplifications and omissions of the so-called elements of lesser importance and of deviations, departures from a strict adherence to the rules. Methodology conceived in such a way in some sense does not allow room for scientific criticism, which would show that actual procedures correspond to rules. Criticism, so frequent in reviews, reports, or in grading assignments, is unjustified from the point of view of methodology restricted to the language introduced by Tarski. Anything that defies a norm simply does not belong to science and cannot be examined in methodology. Things being subjected to criticism, and therefore actually examined in science or in methodology, in this position are outside science and are not amenable to any evaluation. However, since criticism is so important to science and to its development, we ought to take a different position on the issue of names of expressions. We simply have to introduce to the language of methodology names of actually formed expressions, even if they diverge radically from what is allowed to be formed or from what one ought to form in a correct language. Such an augmentation of the language of methodology will make it possible to compare what has been actually effected by the prescribed rules.

Such comparisons are important from the evaluative point of view, which is our concern here. The point is that expressions which it is possible to form according to rules are usually treated in such comparisons as standards or norms, and thus as something which is a value or a bearer of value. Actually effected expressions are compared with these values. It turns out that a comparison of actual results and scientific procedures with the ones which it is possible to effect according to valid rules shows that there is a place for values in methodology. For the time being, we are unable to decide whether such comparisons do occur in science or methodology. Maybe these remarks belong merely to the fringes of science. Anyway, we have shown the possibility of such comparisons – and thus of evaluation.

To realize better what sorts of comparisons we may have to make, let us recall the types of expression-forming rules. Rules of the first type are used in forming and transforming expressions. These are rules of forming propositions and rules of inference, belonging to syntax. Rules of the second type determine, to some extent, the meaning of expressions, rather than their external form. In extensional semantics expressions are permanently way assigned various objects: individuals, sets, classes of sets, ordered pairs, relations, and functions. Expressions belonging to different syntactic categories are assigned different types of objects, accordingly. In intensional semantics, intensions (which shall be covered in more detail below) are permanently assigned to expressions. Intensions in turn assign to expressions the above mentioned objects in a way depending on the circumstances of uttering the mentioned expressions. Objects assigned to intensions belong to types linked with respective categories. So, for example individual names are assigned intensions that, under appropriate circumstances, assign individuals to them while predicates are assigned intensions assigning relations to them, etc.

Actually formed expressions can be compared with their regular standards in all of the above respects; thus, can be compared with the standards in terms of sentence structure correctness; in terms of correctness of inferential transformations; in terms of assignments of sets, relations, etc.; in terms of appropriate correlations of sets, relations and functions with the circumstances of uttering respective expressions. Examples of what counts as errors are syntactic equivocality and vagueness.

2. The second element of the language of methodology consists of translations of the language expressions of the discipline under investigation. Here problems are even more numerous. Different sciences talk about different types of objects and it is doubtful whether general methodology should incorporate translations of statements of all sciences for it is doubtful whether all these translations could form a coherent and consistent language unless we accept, without any particular justification, that the statements of all sciences are reducible to one assumption type, for example to that taken from physics. Reduction of this sort would allow the reduction of languages of particular sciences to the language of, e.g., physics. The state of science today renders such a reduction unlikely and therefore we shall not assume the possibility of carrying it out. As a result however, we face the problem of excessive abundance in the sphere of translations belonging to the language of methodology. This problem can be solved in two ways: 1) We can assume that there is not one methodology but that there are as many methodologies as sciences or types of sciences, or 2) General methodology deals only with what is common to all sciences. Both solutions have their advantages and disadvantages. In the traditional approach methodology declares it aims to investigate what is common to all sciences. In reality, though, the first approach to general methodology is followed. There is a clear distinction into a separate methodology for deductive sciences and a separate methodology for empirical sciences, while the natural sciences and humanities are also treated separately. These three types of methodology have, in some versions, little in common. Let us examine the language of methodology corresponding to the second approach.

The language of methodology ought to incorporate examples of expressions common to the languages of all sciences. It appears that the only such expressions are terms of logic. Even mathematical terms do not appear explicitly in all sciences, although one could potentially find them existing there for languages of all sciences contain expressions like "and," "not," "all," and "is," by means of which all mathematical terms can be defined. It has been accomplished many times, and the most elegant way was probably suggested by W. O. Quine. However, if the only terms common to all disciplines are logical terms, it is doubtful whether a general methodology limited to translating just these terms could distinguish empirical sciences from formal sciences. Terms of general methodology would simply be the same as the terms of methodology of deductive sciences. To emphasize the uniqueness of natural sciences, terms common to all empirical sciences ought to be introduced into the language of methodology. Unfortunately, things are very controversial here. One can adopt, for example, a reistic option as far as languages of methodology and of science in general are concerned. In this case, the language of methodology will incorporate translations of expressions that are, according to reism, common to all disciplines: names of concrete objects and predicates such as is extended, is resistant, is green, is warm, is a conscious body; as well as predicate-forming functors of predicate argument such as: at the time *t*. The assumption of this option is that the remaining language expressions can, in one way or another, be reduced to these expressions and to expressions similar to them by means of definitions. However, not all specialists are willing to endorse the assumption underlying reism. There are problems with the concept of set, which made Quine, a great advocate of nominalism, introduce the concept of set as a primitive and irreducible concept. There are also problems with the concept of possibility and possible worlds, which are difficult to reduce to concepts relating to the actual world. Also, an adherent of reism must show effectively that all the terms are reducible to the terms they use. This is altogether not easy and – most importantly – has not been accomplished so far.

It is difficult to conclude what the terms common to all empirical sciences are – this issue requires a much more thorough investigation. One thing is, however, certain – the language of general methodology ought to incorporate, apart from terms of logic and mathematics, additional translations of expressions common to empirical sciences. To single out the humanities it is probably necessary to introduce additional terms. They might be terms such as "to understand," "to feel" and "to anticipate", and "to intend." The picture of methodology would then look as follows: methodology of deductive sciences would have the most limited domain of translations. Methodology of empirical sciences would incorporate empirical terms in addition to everything comprised in the methodology of deductive sciences. Methodology of humanities would incorporate still more. It is, however, difficult to talk about a unified general methodology; that which is common to all disciplines should rather be called logic of science.

By means of the terms we have introduced above, i.e. names of expressions, translations, and logical terms (which here belong to the domain of translations), it is possible to define the terms of importance to methodology. These are, among others, concepts of: proof, consequence, axiom, thesis, consistency and truth. At least two of these terms are sometimes used as evaluative expressions: the concepts of truth and inconsistency. The term "truth" and its grammatical variations are used as a positive evaluation. The term "inconsistent" is, as a rule, used to disqualify. Do then current terms of methodology incorporate, in a concealed way, evaluations, i.e. terms concerning values? It seems that, in spite of the actual use of these terms in an evaluative way, we cannot assign them such a status now. For now, they are descriptive terms of methodology. The grounds for such a belief are as follows: I assume – without pointing to the relevant opinions taken from general axiology – that what has been evaluated positively and is realizable ought to be realized, and what has been evaluated negatively ought not to be realized, as far as possible. If, therefore, truth is a positive value and if it is achievable then we ought to strive to achieve it. Inconsistency, in turn, ought to be avoided. So far, however, we have introduced no deontic terms concerning duties and obligations to the language of methodology. Therefore we are unable to express, in the present language of methodology, that one ought to strive to achieve truth and avoid inconsistency. So far we can define the concept of truth (or we can introduce it as a primitive term of semantics) and give an account of its properties. For the time being it is functioning in a purely descriptive way. Also, our language does not contain evaluative terms like right, wrong, etc. It is easy to notice that our present considerations are conducted on a metamethodological level – we are deliberating about methodology and hence we use the concepts of right and wrong, of obligation, etc. When, therefore, truth and inconsistency work in an evaluative way, it probably takes place not in methodology but in metamethodology or, possibly, in protoscientific considerations. Is metamethodology a separate domain of inquiry or does it belong to methodology in a wider sense and is the lower level of the last methodology? That is the question.

3. The third component of the language of methodology in Tarski's interpretation are the specific terms of methodology, which go beyond what is dealt with in a discipline under investigation. In Tarski it was only one term, namely, the concept of concatenation. Concatenation serves the purpose of forming names of complex expressions. For example if α and β are names of expressions: "John" and "walks," then α followed by β is the name of the sentence "John walks." The expression "followed by" designates concatenation: placing one expression after another. General methodology requires many such specific terms.

It has already been mentioned that we can distinguish empirical sciences from deductive sciences if we introduce into our methodology the names of objects and predicates like: is extended, is green, is resistant, etc. Such terms do not enable us, however, to stress the uniqueness of justification in empirical sciences. They enable us only to limit the subject of inquiry of deductive sciences to the subject of inquiry of empirical sciences. In deductive sciences justification consists in showing that a formula is an axiom or is derivable from axioms by means of applicable transformations. In empirical sciences we do not treat any propositions as axioms. Even so-called perceptual statements can sometimes be revoked. If, however, perceptual statements were irrevocable (as the Vienna Positivists maintained) they still could not be conceived of as axioms. That is, we could not assume that they were true because they are mutually noncontradictory, because they make up a complete or categorical system, and so on. Perceptual statements are accepted as a basis for accepting or rejecting other statements because they are based on observation; that is on certain cognitive operations and their results. So, therefore, general methodology should be enriched with expressions referring to cognitive operations and their results. There are a number of such operations. They include not only watching, listening, touching, etc., but also applying measuring devices to the objects being measured or effecting specific changes under controlled conditions. Moreover, introducing such expressions does not guarantee that it will be possible to use them to distinguish humanities from natural sciences, with respect to the method of justification. The issue lies in determining whether the statements concerning understanding, anticipation, feeling, etc., are justified only by means of operations, such as watching, listening, touching, etc. If it were so, behaviourism would be fully justified. The only thing we can say today is, however, that behaviourism was at most an ambitious research programme. Since the justification of statements about feelings cannot be, as far as we know today, based completely on the so-called external observation, we have to add the specific terms of methodology of humanities expressions that refer to cognitive operations characteristic of humanities. We mean especially the procedure for understanding texts, conversations, questionnaires, etc. This could also include understanding human beings: ascribing to them, by means of a certain analogy, properties which observers ascribe to themselves under given circumstances.

Cognitive operations, whose accounts we are to introduce into the language of methodology, are either actually performed or are feasible – in particular, feasible to be performed according to some rules or clues. We should, then, divide operations into two groups: actually performed operations and potential operations. This division is analogous to the division of expressions. It differs in some respects, however, from the division of expressions. In artificial languages (common in various scientific disciplines) rules of expression forming are stated quite explicitly. In natural languages these rules are intuitive. Despite this, we usually know whether an expression is well-formed or not. In the case of operations rules are usually intuitive but we are unable to determine whether an operation is performed correctly more often than in the case of forming expressions. Despite a lack of clearly stated rules of performing cognitive operations we nevertheless do compare actually performed operations with their more or less known standards. When an instructor criticizes and instructs her student with respect to his or her cognitive operations, she must have some standard in her mind. This sort of intuitiveness of our knowledge has been described accurately by Fleck.

As in the case of comparing expressions formed actually with potential or standard expressions, in the case of cognitive operations we may come across evaluative elements: potential cognitive operations may be conceived of as standards or as paradigms. As in the previous case, we cannot, however, assume for sure that in methodology actual cognitive operations are compared to and evaluated from the point of view of standard values. The reasons to doubt this are the same as before. So far we do not have any evaluative or normative terms in the language of methodology. The question of whether we should have any such terms in methodology is a different issue.

We constantly bear in mind the applicability of the language of methodology to giving an account of justification and to describing its properties. This issue surfaced for the first time when we extended the language of methodology to enable us to distinguish justification in deductive sciences from justification in other sciences and in the humanities. At this point we have to remember the need for an adequate account of justification.

In deductive sciences a proposition is justified (proved) or unjustified irrespective of whether actual proof has been found and presented. This is the notion of potential proof. Things are different in natural sciences. The notion of potential proof or, rather, potential justification is less important there. If we assume that basic propositions, playing at least some role in rejecting misguided ideas, are essential components of justification, then we shall see clearly that the notion of potential justification is of little use in natural sciences. Since the times of Popper's *Logik der Forschung* it has been almost generally assumed that basic propositions are revocable. As our cognitive abilities increase, what used to be considered accurate becomes inaccurate, what used to be a fact turns out to be an artifact. Finally, and most importantly, existing theories are totally reinterpreted in the light of new theories. These changes are much more important to the natural sciences than determining what the currently accepted statements entail. Obviously, in some cases it is necessary to investigate what the consequences of accepted hypotheses are – at least to reject them. It is, however, the more important changing observational data, mainly an inflow of new data that makes a theory an outdated hypothesis. What is justified in the light of one set of data and basic propositions might be unjustified or rejected in the light of a new, for example bigger or more accurate, set of data, and hence of basic propositions. When Aristotle deliberated whether the Earth revolves around the Sun, he supposed that he should observe some movement of nearer stars against the background of farther ones. He observed no such movement and therefore he concluded that the heliocentric hypothesis was falsified by his available observational data. He had no data concerning the size of our galaxy, either. When we consider this issue nowadays we know not only that there is such movement but we also know the source of Aristotle's mistake: his lack of knowledge of the size of our galaxy and the imprecision of observational instruments used by the Greeks. Thus, justification of propositions depends simultaneously on observational capacities and on the theories accepted. It also depends on researchers' interests. If Fleming had ignored the fact that a bacterial culture placed near a mould culture had died out, we would probably not have antibiotics today. In brief, justification should be considered in relation to the set of basic propositions accepted at a given time. This set, in turn, depends on instruments, interests, accepted theories and even on available funds, etc.

As it will later turn out, relativization of justification to circumstances, some of which have been mentioned above, entails relativization of propositions, or speaking more strictly, of their semantic referents to these circumstances. In brief it will suffice to mention that justification of some hypothesis by means of basic propositions can change the interpretation of fundamental propositions: changing the hypothesis could be followed by changing the semantic model for interpreting the propositions. Thus, referents of propositions can change. Thus, referents or denotations of a discipline under investigation depend on the circumstances of justification. Such dependence of extension (a general name for referents and denotations) on circumstances is characteristic of so-called pragmatic-intensional semantics, introduced by R. Montague.

In methodology, we ought to be able to describe dependence of justification on changing circumstances as well as what follows from it, namely dependence of extension on circumstances. Hence we should consider what the language of methodology ought to be like to fulfil this condition. We are unable to do this because the language of methodology contains – so far – only the names of expressions and translations of expressions of the disciplines under investigation. In this language we can describe only the assignments of extensions to expressions that are lasting and independent of circumstances. We simply have to introduce to the language of methodology the names of the circumstances of justification and of the circumstances influencing the changes of extensions of the language of the discipline under investigation. R. Montague and D. Lewis, who paid much attention to the circumstances influencing changes of extensions, do not list all circumstances. When Creswell talked about such circumstances he generally counted as circumstances all the properties of expressions. This suggestion is less appropriate here because, for example, an ability to conduct observations is not a property of propositions but of the people carrying out observations. Montague and Lewis did not pay attention to the ability to conduct observations, research equipment, or the sets of fundamental propositions, etc. They were so fascinated by the possibility of explaining properties of modal and occasional expressions that they did not notice the circumstances influencing justification and thus influencing the extensions of propositions. They limited their investigations to circumstances like the time and place of speaking, the sender and receiver of signs, the gesture of pointing, and to some others as well. It is impossible to list all the circumstances influencing justification of hypotheses; we just postulate that they should find their place in the language of methodology as a second group of its specific expressions.

There are statements of clearly evaluative character referring directly to objects under investigation. One says, for example, that an issue is interesting, that some regularity is crucial to solving some wider set of problems; one talks about marginal and essential phenomena. Such kinds of statements can be found in texts on both natural sciences and humanities. In the latter they are of course much more abundant. A historian of music can say, without evoking readers' protests, that Bach's music is monumental and a historian of art can describe a Gothic cathedral as a wonderful work of art. Qualification of these statements, however, raises some doubt. First, one can think that they are not authentic statements of a given discipline and that they do not belong to its first-order language. One could classify them as protoscientific propositions. In the case of natural and mathematical sciences argumentation in support of this possibility is very simple. Laws of these sciences do not contain any evaluative terms like the ones we mentioned above (interesting, important, essential, etc.). It is more difficult to put things in this way in the case of humanities. Bearing this possibility in mind, let us consider another possibility. We will choose a better solution later. These terms can be treated as parts of the metalanguage or of the language of methodology. As we have said, they do not occur, at least in natural sciences, in laws and theoretical hypotheses *explicitly* formed. They occur in the propositions judging whether a subject is worth investigating, whether research is worth continuing and whether research ought to be conducted in the same way as before. All these issues belong to methodology. Moreover, these statements can be modified: they can obtain a metalinguistic form and then be regarded not as referring to objects but to propositions. Instead of talking about important regularities we can talk about important laws. If these terms belonged to the language of methodology they would belong to the domain of its specific expressions.

Still another approach totally eliminates evaluative terms both from the language of science and from the metalanguage. Here they are considered simply as terms expressing researchers' subjective attitudes, which, from time to time, express themselves in an unconstrained way. These terms would be expressions of a psychological and autobiographical nature. This approach, however, suffers from a predicament. Evaluative terms are common in scientific discussions and it would be hard to imagine them without such terms. The same is true for reviews, assessments, etc. Excluding these terms would also make it impossible to take part in discussions, to write reviews and to perform other similar procedures, important to conducting scientific investigation.

On the other hand there is an approach according to which terms evaluating objects of investigation are genuine members of the first-order language of science and of the translation domain of the language of methodology. This approach is favored by humanists. They investigate, among other things, people's evaluations, so evaluations are subjects of investigation. Here a humanist does, of course, talk about evaluations and the translation domain of his methodology must accommodate appropriate translations. In natural sciences the situation is less clear but even there we can find evaluative elements in disguise. It is common that we never investigate objects as they are in their ontological completeness. We always make simplifications and idealizations. We leave out less important properties and consider only the more essential ones. Sometimes the most important properties are prominent; sometimes they are obscure. An evaluation of their importance frequently depends on accepted theories, needs, etc. At any rate, it is possible that in natural sciences we do make evaluations and that evaluations not only reveal our attitudes toward objects but also, occasionally, significantly influence the course of investigations.

This paragraph, together with the previous one, highlights the points in which one can find in methodology evaluations and expressions that mention the value of subjects of investigations. Showing these possibilities, though, is not proof that evaluations are actually permissible in methodology. We will try to give answers to questions arising in connection with this subject while considering issues involving not language but justification.

Justification

1. Due to the existence of different methods of justification in different disciplines and at different stages in the development of science we will adopt a very general understanding of the term "justification." First, we will assume that the aim of justification consists in determining, first of all, true propositions. If this cannot be achieved (some theorists question the possibility of determining truth), then it involves detecting probable propositions. If this cannot be achieved either, as Popper maintains, then the aim of justification consists in discovering verisimilitudal propositions, and if this proves to exceed our abilities we have to accept propositions that cannot be disproved at a given stage in the development of science. If neither of these requirements can be achieved then justification aims to discern propositions that might turn out to be useful in practice. The methods for achieving these goals are constrained only by one condition: the course of justification ought to be publicly controllable by a group of professionals that the whole community accepts as competent in a given domain. The result of justification consists in accepting a proposition as true, probable, verisimilitudal or useful.

These remarks on justification are controversial but they are not the main battleground in the discussion of justification; they are protected by being too general. The fiercest discussions are spurred by investigations into what procedures count as justification and which of them fulfil the conditions of intersubjectivity, enabling us to achieve truth or some semblance of it.

2. There are two simple observations underlying our beliefs concerning the worthiness of methods. Firstly, we are less likely to be wrong in cases that are simple and when they are not too numerous. Overabundance and complexity of cases under investigation confuse us with too many details and an even greater number of relations. Precisely because of this Descartes recommended beginning with simple issues and building on this foundation, systematically and consciously, a knowledge of complex issues. Secondly, if there are no limits to our imagination and if nothing constrains it different people attain different visions. This though causes discrepancy – for not more than one vision can be true – but they can all be wrong. If our cognition is captured by a compelling reality or by a structure of mind there is, at least, no discrepancy and we can, at least, talk about intersubjectivity. It is commonly held that sensory perception is forced by reality and that the role of imagination is minimal in this domain. It is maintained, less commonly, that reality directly determines our intellectual knowledge, resulting in an intellectual insight into the world. Here, intellectual evidence is involved. It is held, still less frequently, that the structure of our mind – common to all people – determines a common, intersubjective apprehension of the world. The following deals with the first approach.

Simple sensory perception was thought of (e.g. by the Vienna Circle) as being free from error if the observations were performed by adequately trained people under appropriate conditions and if the observations could be controlled by repeating them. To avoid errors in the case of abstract knowledge (related more closely to our imagination), the Vienna Circle understood justification as the reduction of all knowledge to sensory perception or as grounding all knowledge in sensory perception. It soon turned out that it is not possible to reduce all scientific knowledge to sensory perception. It very quickly turned out that theoretical terms cannot be clearly defined (by means of normal definitions) in so-called perceptual terms. Moreover, this cannot be achieved even by means of partial definitions. And it is logically not possible to deduce general propositions from basic propositions. As such a simple reduction turned out to be impossible, there followed trials to ground scientific knowledge in basic propositions in a different way. Basic propositions were also called perceptual statements, basic statements or protocolar statements but we need not consider the differences between them. It was assumed that basic propositions could be used to rule out hypotheses that are evidently false. Such a justification is obviously much weaker for there is no guarantee that the statements that are not yet evidently false and might be accepted at a given moment, will turn out to be false after a set of data changes, e.g. expands. Our historical experience teaches us that, as a rule, even very well grounded theories are rejected after some time.

3. For a hypothesis to be rejected on the grounds of basic propositions it is necessary that basic propositions follow logically from the hypothesis, from basic propositions describing empirical initial conditions, and from auxiliary hypotheses. If basic propositions turn out to be false because they contradict observations, the hypothesis is to be rejected if auxiliary hypotheses cannot be rejected. This general scheme of falsification, which was put forward by Popper, has several interesting features. Firstly, it is known that a given set of propositions

Q (which can be arbitrary) can follow from many different sets of propositions. In other words, a given set of basic propositions can follow from a great many different explanatory hypotheses. Basic propositions do not determine unequivocally explanatory theories or hypotheses. Accordingly, a given set of basic propositions describing observations available at a given time can be explained in different ways and there arises peculiar competition among the explanations. Such a situation is not troublesome if, as a result of acquiring new facts, all hypotheses but one are falsified. Then the remaining hypothesis (usually a conjunction of propositions) becomes an accepted explanatory theory.

But if neither of the competing theories is fully satisfactory the situation begins to become troublesome. The old theories have well-known deficiencies while the new hypotheses are not fully understood and they seem to contradict accepted approaches all the time. Both types of competing hypotheses cannot explain everything; moreover, it is not possible to describe by means of their languages everything that seems interesting to us. We might as well think that all theories are equally falsified. It is, however, philosophically possible to suspend acceptance of all those hypotheses and wait for the verdict of the strictest of judges: time itself. However, life and practical needs do not allow such a philosophical approach. Decisions must be made but, at the moment, the basic propositions are not a sufficient basis upon which to reject some hypotheses and accept just one. Waiting to see which hypotheses represent a so-called degenerating research programme and which a progressive program, which keeps delivering new facts, requires time as well. Therefore Lakatos' concept of research programmes is not the best solution to the difficulties we are facing. Approaches derived from Popper's falsificationism do not solve the problem of many competing hypotheses although they do ordain to expose a hypothesis to strict tests, including extensive discussions. The source of such a state of affairs is that during the evaluation of a hypothesis they take into account only its relation to basic propositions.

4. Difficulties in giving an account of justification of a hypothesis in terms of its relation to basic propositions causes a problem: what other additional factors can influence its justification? Since justification influences the uniformity of acceptance of a hypothesis or gives some security that a hypothesis might be true, probable, or verisimilitudal, these additional factors must influence these properties of the hypothesis. There are many such factors. We ought to remember that they do not replace the relationship a hypothesis has with its basic propositions. They simply add an important element of justification.

When we investigate the justification of a hypothesis, we, as a rule, accept a number of other theories. One of the criteria of merits of a hypothesis is the possibility of it conforming to accepted theories. If it contradicts them then there arises a dilemma: Are we to accept the new hypothesis or are we to reject one of the old theories? The conformity of a hypothesis to the established body of knowledge facilitates its acceptance, for theoretical constructs support one another in the case of conformity or destroy one another if they are mutually contradictory. As we have said, such conformity is not a decisive element of justification. It is just one of many elements and only a full set of them brings about a final decision. Among previously accepted theories we can find, among others, accepted philosophical beliefs.

While justifying a hypothesis, we make use of different methods: inferences, observations and analyses of different kinds. Usually these methods have already been tried. But one can, however, question whether they can be applied to a new hypothesis. Speaking more generally, one can question whether a given way of arguing is adequate. This criterion has a methodological character but it is difficult to imagine that one can justify hypotheses in science without any reflection on methods. An investigation of the methods that one uses is, then, a part of justification.

After a hypothesis has been discussed thoroughly and openly we treat it in a way different from the way we had treated it before it underwent public criticism. The concept of a strict test of a hypothesis embraces, first of all, presentation of it to many scientists to review and compare it with other ideas, finding as many of its consequences as possible. It also involves determining the premises of the hypothesis, which are accepted, often unconsciously, by its authors. The possibility of discussing a hypothesis assumes that it is formulated in an appropriate way. It must be presented in such a way that competent researchers can understand it without effort. Firstly, a hypothesis must be complete as to what it regards. It must be precise enough to enable determination of how it is different to other ideas. A hypothesis should also allow conclusions to be derived from it. Deriving conclusions is a purely logical procedure and should not cause the authors of the hypothesis to protest that their hypothesis is being altered and that the conclusions somebody derives do not follow from it. If such protests are frequent then there arises a suspicion that the hypothesis is formulated in a wrong way and that it allows (almost) arbitrary interpretations. In brief, the actual formulation of a hypothesis is one of the decisive factors in its justification, regardless of Popper's postulations about precision.

Conformity of a hypothesis to existing theories is, as we have noticed, a foundation for treating it in a positive way. It happens, however, that hypotheses that contradict existing theories are preferred. A typical example of a hypothesis which had been falsified and which contradicted accepted views is the heliocentric hypothesis of Copernicus in its original form. After Tycho de Brahe's observations it was essentially falsified. Ptolemaic theory modified by Tycho de Brache explained and predicted the facts better, while Copernican theory sharply contradicted the established views concerning the dimensions of the world and its structure. In spite of this, it was not forgotten; it remained a matter of interest despite its evident deficiencies. Only after Kepler's corrections was it possible to discuss the advantages of the heliocentric theory over the geocentric theory. What was the reason why it was not condemned by the community of scientists? It may have been the fact that it was original and that it challenged almost everything accepted at that time. The Europe of the day simply wanted changes and innovations – almost *per fas et nefas*. As a result of this desire, Copernican theory was considered for so long that it was eventually corrected. Accordingly, the desire has essentially influenced the final acceptance and justification of the theory.

Practical usefulness is one of the important circumstances influencing acceptance of a hypothesis. This feature is related to the informativeness of a hypothesis. A hypothesis that is too general is, as there are alternative interpretations of the ways of understanding it, less prone to falsification. It is *a priori* more likely to be true. In spite of this, we would rather accept hypotheses with a richer content or more precise ones, as long as they have not been falsified. Informativeness prompts us to accept a hypothesis regardless of subjective, cultural, etc. attitudes. In brief, it is a component of justification.

Apart from these elements of justification, there are also others, which are not appreciated now, although they are actually taken into consideration. Among them is authority. Formerly, it was a very important factor of justification; nowadays it is apparently passed over in diffident silence. Nevertheless, what a well-known and respected specialist says does have some weight to it and if, for a lack of other data, we have to consider opinions of specialists in a given field, then authority still counts. Also, involvement of a powerful organization influences the acceptance of a hypothesis. These factors are listed here in order of increasingly psychological and sociological character. However, they are not subjective. If there are no other reasons, it is more reasonable to side with authority. It is also reasonable to suppose that a more popular view might be more likely to be true or that it is more probable. It is so, of course, if there are no other data available. The above criteria of justification are bankrupt because people tried to present them as opposite to other types of justification. But if they serve only to augment an argument or to amend its deficiencies, they are perfectly rational and appropriate.

It is difficult to give a full account of all circumstances that intersubjectively influence acceptance of a hypothesis; there are probably many more of them than have been mentioned above. It is immediately possible to find other such circumstances while analyzing and discussing issues of justification and of proposition acceptance. Different factors can come into play under given circumstances and, what is most important, they are mutually incompatible. We have already given some examples of such incompatibility. Informativeness stands in opposition to probability; conformity to accepted theories contradicts originality and novelty; authority is opposed to rigorousness of tests, etc. In such situations it is constantly necessary to consider which factor has, in a given situation, a greater capacity to justify. Sometimes it is necessary simply to make decisions under the circumstances of obvious uncertainty. Worse still, we are often unable to determine – as we have mentioned above – all the elements of justification.

Values

1. As we have already said, justification can be conceived of as choosing some propositions and their acceptance on the grounds of their relationship to given basic propositions; especially to theories, opinions of authorities, opinions of institutions, or also on grounds of their informativeness, originality, appropriate formulation, and of methods of finding basic propositions being a basis for propositions we choose. These circumstances have various degrees of influence on the acceptance of propositions, and this influence has to be uncovered during discussions, or has to be simply divined. If circumstances are mutually conflicting, we must make a choice. We will, of course, consider the factor that we think has the most influence on the acceptance of propositions under consideration. We shall assign to these factors degrees of influencing justification. These degrees can vary depending on additional circumstances. Moreover, on the basis of the known circumstances of justification, we can accept a hypothesis to varying degrees; we can also not accept it or even reject it.

Let us try to summarize our considerations. We shall list the degrees of acceptance in the column, from the fullest acceptance to complete rejection. Factors influencing acceptance of a hypothesis *p* shall be listed in rows. Because different factors influence acceptance in a different way, intersections of columns and rows will hold magnitudes of these influences. Let us assume that we assign the magnitude of 10 to basic propositions and that of one to authority. This means that basic propositions influence acceptance of a hypothesis ten times more than the authority of the author of the hypothesis. We shall not consider actual cases; we are interested only in giving a general idea, so instead of numbers we shall use numeric variables. To make things simpler, let us assume that the variables take values only from the set of natural numbers. To fit the table on a page, we shall represent factors influencing acceptance with capital letters in the following way:

A – sets of basic propositions,
B – their logical relations to the hypothesis *p*,
C – sets of accepted theories,
D – their relations to the hypothesis *p*,
E – sets of observational and experimental procedures,
F – the language of the hypothesis and rules of inference,
G – informativeness of the hypothesis,
H – preciseness of the hypothesis,
I – originality of the hypothesis,
J – authority of the author of the hypothesis,
etc.

	A	B	C	D	E	F	G	H	I	J	–
Acceptance to the degree of one	i_1	j_1	k_1	l_1	m_1	n_1	o_1	p_1	q_1	r_1	–
Acceptance to the degree of a/b	i_2	j_2	k_2	l_2	m_2	n_2	o_2	p_2	q_2	r_2	–
Acceptance to the degree of c/d	.	.	.	.	.	.	.	.	.	.	–
.	.	.	.	.	.	.	.	.	.	.	–
.	.	.	.	.	.	.	.	.	.	.	–
.	.	.	.	.	.	.	.	.	.	.	–
.	.	.	.	.	.	.	.	.	.	.	–
Acceptance to the degree of zero	.	.	.	.	.	.	.	.	.	.	–

By justification we mean a choice of acceptance and to what degree, with respect to factors A, B, ... and to the magnitudes assigned to them. There are many types of justification, which is due to the following possibilities. Circumstances A, B, ... can change. Circumstances A do not form a single set of basic propositions but they form a different set at each stage in the development of science. We can assign different degrees of influence on hypothesis justification to different sets of basic propositions. But even if both A and the remaining circumstances are constant, different specialists might assign different degrees of influence on hypothesis acceptance to the same circumstances. It might turn out that the final outcome is the same in all cases, i.e. that the degrees of hypothesis acceptance are the same. In spite of this, however, the justification would be different. For by justification we mean a function of the following arguments: circumstances A, B, ..., and their magnitudes; the value (result) of the function is acceptance to a certain degree.

2. The existence of a large number of functions called justification suggests that they are a subject of choice, too. We shall call a specific choice, or rather a type of choice, a scientific strategy. When determining our scientific strategy we choose for some reason, a given type of justification out of the many possible ones. It seems that the choice of a given type of justification is influenced by factors different from those that influence the choice of degree of acceptance. If we aim just to convince somebody, are not interested in applications or are not interested in the truth, then we take our types of justification from rhetoric and from similar domains. If we aim to convince somebody of a hypothesis that is to influence his practical conduct, then justification will rest not only on the rules of classical logic but also on the methods dealt with by decision theory and game theory. Using rhetorical arguments in mathematics would be very strange. Shortly, a particular choice of justification depends, among other things, on our goals, which can be truth, probability, convincing somebody, or influencing a decision to act. We use different types of justification if we lack time (for example during a battle) than if we have an indefinite amount of spare time to devote to investigation. A choice of a particular type of justification is also influenced in an obvious way by the technology available to us. A nineteenth century physician would justify a thesis that somebody suffered from tuberculosis differently from the way we would today. Similarly, a choice of method is influenced also by financial considerations but, first of all, by the results of the cognitive activities conducted so far, by our being used to them and by our conviction of their usefulness under given conditions. A general understanding of the way in which bodies behave is one of the components of evaluating what types of justification can be used. It is not possible to list all the factors influencing a choice of justification, just as in the case of choosing a type of acceptance. For our purposes, it is enough to know that such factors exist and that each of them is a whole set of possibilities and that, therefore, there exists a whole set of scientific strategies.

A scientific strategy, or a choice of a type of justification, can be presented in a table, just like a choice of acceptance or justification.

A' – an expected result: truth, common beliefs, etc.,
B' – a picture of the world allowing for the intelligibility of the world,
C' – technical resources for conducting cognitive procedures,
D' – expected usefulness of knowledge,
E' – cognitive experiences gained so far
F' – expected difficulties and cognitive effort
etc.

	A'	B'	C'	D'	E'	F'	–
Justification 1	i_1	j_1	k_1	l_1	m_1	n_1	–
Justification 2	i_2	j_2	k_2	l_2	m_2	n_2	–
Justification 3	i_3	j_3	k_3	l_3	m_3	n_3	–
.	.	.	.	.	.	.	–
.	.	.	.	.	.	.	–
.	.	.	.	.	.	.	–
Justification n	i_n	j_n	k_n	l_n	m_n	n_n	–

3. According to what we said at the beginning of the present article, factors influencing decisions to act are values or factors that bear values. Numbers at intersections of rows and columns are measures of these values. The procedures described in these tables are taken from decision theory. Sometimes, however, objections arise. For how can a set of basic propositions be a value? In spite of this objection, it intuitively makes sense to treat a set of basic propositions as a value. If we are to justify a hypothesis (i.e., to determine what might contribute to its acceptance), it is not indifferent to us what the available set of basic propositions is. We want to be in possession of a set of basic propositions as large and as diverse as possible and are pleased to have such a set at our disposal. We treat a set of basic propositions as a real value. If the theories we accept are to confirm a hypothesis being investigated by virtue of their conforming to it, we also treat them as values. If the theories we accept are to confirm a hypothesis being investigated by virtue of being in agreement with it, we treat them as values, too. If a hypothesis agrees with, say, existentialism, then a critic of the hypothesis decries it by saying: "But that's mere existentialism!" Such a reaction clearly shows that we treat theories as values.

Thus, we have reached the conclusion that methodology, which deals with justification and choices of justification, i.e., with scientific strategies, must deal with values. Justification and strategies are choices made on the basis of the values ascribed to the circumstances of actions. What we thought was a mere possibility turned out to be a fact. Now, we can see quite clearly that methodology does use standards. If we ascribe a positive value to cognitive procedures that have so far given positive results, then new procedures will be assigned a positive value depending on their similarity to the established ones. Obviously, showing a general outline of a scientific procedure does not solve the majority of problems. Let it suffice to say that we do not really know the basis on which we assign magnitudes of value to circumstances. Solving these problems seems to be the main task of today's methodology.

Relativism of methodology

1. At the beginning of the present article we remarked that the issue of values in science is crucial to the controversy of whether scientific knowledge is sound or temporary, relative or subjective. One thing is clear: acceptance of hypotheses and the choice of the way of justifying them depends on many conditions. Not all of these conditions are known. Scientific knowledge depends, in turn, on them. This type of relativism does not, however, threaten the belief that scientific knowledge is sound and nonsubjective. There are also relativisms of a different sort, totally questioning the value of scientific knowledge. According to such views, science is almost a silly game played by people who want to be taken as serious scientists.

According to the later ideas of Carnap and Braithwaite, terms occurring in basic propositions do not define theoretical terms occurring in abstract explanatory hypotheses. It is the opposite; theoretical terms are used to define perceptual ones. Independently of such definitions of theoretical terms, perceptual terms have their additional characterizations in ostensive definitions. Let us assume now that we are to explain a given set of data (and that data are expressed in basic propositions). Several mutually contradictory theoretical hypotheses are offered as its explanations, and since they are contradictory they must be interpreted in different models. Terms occurring in an explanatory hypothesis specify terms occurring in basic propositions so, therefore, a change of interpretation of theoretical hypotheses entails a change of interpretation of the basic propositions in the model. Interpretation of basic propositions depends then on the interpretation of hypotheses and also on which of the competing hypotheses is accepted. Hypothesis acceptance depends in turn on various circumstances, e.g. on the size of the set of basic propositions, methods of conducting observations, on agreeing with other theories, etc. Therefore, interpretations of all the propositions of the theory, including basic propositions, depend on all these circumstances. Interpretation of terms consists in assigning the individuals, sets, relations, and functions of the various types to them. They are collectively called extensions. It turns out that extensions of expressions depend on these circumstances. Such a result is in perfect agreement with Richard Montague's idea of pragmatics in which extensions depend on the circumstances of uttering expressions: time, place, and speakers and receivers. We have gone a bit further. Extensions depend also on the circumstances influencing proposition acceptance and, indirectly, on the circumstances influencing choices of justification. But the underlying mechanism relating extensions to circumstances is the same. A change of an extension does not influence the meaning of expressions if we consider it to be the same as the so-called intension. Intension is a function assigning appropriate extensions to the circumstances of uttering and accepting expressions. According to Montague, an intension is permanently assigned to an expression. In fact, things can be more complicated than Montague assumed. One thing is important for our investigations. A change of hypothesis and of model does not have to be followed by a change of meaning if the models belong to a class of accessible models. We can assume that models are readily accessible even though they are employed in the interpretation of mutually contradictory hypotheses. They are accessible in the respect that all of them can be employed to interpret given basic propositions which must be true in all these models.

Relativism has thus been generalized. Not only do acceptance and justification depend on circumstances but so do the extensions of the terms and propositions, too. However, this does not threaten the value of science. First of all, scientific terms maintain the same meaning, even if they occur in mutually contradictory hypotheses. So it is not possible to maintain that inconsistent hypotheses form isolated languages that are mutually untranslatable. It is not possible to show that proponents of contradictory hypotheses speak different languages and that they are unable to communicate.

2. Cognitive and methodological relativism, in questioning the value of science, not only relates knowledge to numerous circumstances but also: a) contends that these circumstances cannot be known or, alternatively, contends that the conditions under which hypotheses are accepted are unique and that there is no way to give a general rule correlating knowledge with the circumstances of acquiring it; b) the circumstances influencing hypothesis acceptance can be conceived of as incommensurable, which leads to the same conclusion as before; there is no way to give a general rule correlating hypothesis acceptance and its conditions.

If we relate proposition acceptance to circumstances that are, in one way or another, impossible to grasp, we are unable to describe the course and factors of acceptance. The same is true for justification but then we cannot practice methodology which – as we have said before – investigates justification in science. Someone who assents to radical relativism cannot pursue methodology. They simply do not allow for its existence. So it follows that a methodologist, i.e. a person engaged upon methodology, cannot, under the sanction of inconsistency, assert that radical relativism is true. Listing, categorizing and classifying circumstances influencing acceptance and justification leads immediately to nonradical relativism. Determining the degrees of influence of these circumstances on acceptance shows their evaluative character. The task is complicated but it is not beyond human capacities.

Department of Methodology
Philosophical Faculty
Catholic University of Lublin
Al. Racławickie 14
20-031 Lublin
Poland

Poznań Studies in the Philosophy of the Sciences and the Humanities
2000, Vol. 72, pp. 37-60

Leon Koj

SCIENCE AS SYSTEM

ABSTRACT. Science has often been treated as a system and this last notion is derived from the methodology of deductive systems, after minor amendments have been introduced into the methodology of science. In the paper the Bertalanffyan concept of system has been applied to science (and the humanities). In particular, science is described as an instrumental system that has been appropriately defined. In the framework of this Bertalanffyan conceptual apparatus, divergent aspects of science such as: its informational content, its refe rence, psychological and sociological attitudes towards science, political implications of science, etc. can be coherently discussed. Furthermore, these aspects can be compared and evaluated.

1. General remarks about systems

a. Considerations concerning science (in the broad sense of the word, including the humanities) most often do not use the notion of system as introduced by Bertalanffy. This concept of system will be used in the following discussion.

b. M. Bunge and J.M. Bochenski proposed two different general notions of system. We take into account the one put forward by M. Bunge: system is a system in so far as it differs from its background, thus this background, here called environment, must be introduced into the definition of system. On the other hand, Bochenski points to deductive systems (now treated as Bertalanffyan systems) as having no environment. It is difficult for me to accept Bochenski's statement because without any theoretical background we are not able to discriminate a deductive system, from, say, an infinitely long tape with strange ornaments. The ability to discriminate both entities is due to the pragmatic background of the deductive system. There is also the semantic aspect of the environment. Only the consistent sets of sentences make up a deductive system. Very often consistency is checked with the help of semantic methods. Semantics therefore constitutes the semantic part of the environment of the deductive system.

We assume that the system is a whole composed out of parts which are suitably related to each other. Let $c, c_1, c_2, ...,$ be parts of the whole C. Let $R_1, R_2, ..., R_k$ be the relations between the parts. In turn, let $R_{k+m}, ..., R_{k+m}$ be the relations between the adduced c's or the whole C and the environment U (*Umwelt* in German). In view of these conventions we arrive at the first, simplest, concept of the system:

1. $S = \langle C, R_1, ...R_k, R_{k+1}, ...R_{k+m}, U\rangle$

c. The concept just introduced is rather uninteresting for it describes an unchangeable structure. Perhaps it is useful in describing deductive systems which are given at the start as wholes, even if we do not know its infinite depth. But in the case of empirical sciences the question is quite different. Almost all empirical systems, e.g. organisms, change with time. This holds true of both sciences and humanities. Some new concepts are introduced, some become obsolete and are left out (e.g. phlogiston, the ether), some notions are redefined (e.g. the atom), new mathematical apparatus becomes useful, even the rules of inference can be changed. Even the environment becomes different. It evolves spontaneously and, what is more important, we apply our scientific knowledge and change the surrounding world and change ourselves, by committing ourselves to new convictions. Just as in the well known case of organisms in sciences and theories in different moments, parts of the scientific system are different although the whole remains the same system. Let i be a period of time of indeterminate length. In different i we get different C's, different $R_1,...,R_k$, different $R_{k+1},...,R_{k+m}$ and lastly different U's. The system which lasts many moments i consists of all systems which are given in the different moments i. The moments i are ordered in a suitable way and the ordering relation is *later* or *earlier*. Thus we get:

2. $S^{\{i\}} = \{\langle\{C\}^{\{i\}}, \{R_1, ..., R_k\}^{\{i\}}, \{R_{k+1}, ..., R_{k+m}\}^{\{i\}}, \{U^{\{i\}}\}\rangle\}$

It is obvious that the different systems existing in the moments i following one after another must be in some way connected. The later ones depend on the earlier ones. It seems that the problem of this dependency is the one which is discussed most often. We do not want to go into this wide field of questions, arguments etc. Let us only discriminate between the relations that hold between the successive C^i and other relations. Let the relations just mentioned be symbolised by T. Now, the relation which links the successive sets $\{R_1, ..., R_k\}^i$ is W. Let Y be the relation that connects the successive sets $\{R_{k+1}, ..., R_{k+m}\}^i$. Let the different U^i be linked by the relation Z. The relations T, W, Y and

Z also change with passing time, but to avoid too much complication let us not take into account this new problem.

d. Systems that change and yet do not lose their identity, as a rule, exhibit feed back mechanisms. These in turn create homeostasis which enables the system to exist for some time. Some kinds of homeostasis make it possible to develop the systems that exhibit this homeostasis. The distinguishing feature of evolving systems is that they depend on their environment. They cannot evolve without obtaining energy and materials from the environment. So far we have not discovered any kind of perpetuum mobile. Similarly, all systems which devour parts of the environment without any discrimination cannot survive for long. Both abilities: to take from the environment and to filter out some ingredients from the surroundings are vital for the existence of evolving systems. The lack of the first capacity results in stagnation, the other, in repletion. In the case of theories (and because of that also in the sciences) in the first case we get simply stagnation, in the second inconsistency. Systems which do not filter out fragments of the surroundings cease to differ from it. This is exactly the case when an organism dies. These vague remarks point to the great importance of the relations $R_{k+1}, ..., R_{k+m}$.

Now, we get the last description of system which is important to us; further details will be ignored.

3. $S^{[i \in I]} = \{\langle \{C\}^{[i]}, [R_1, ..., R_k\}^{[i]} \{R_{k+1}, ..., R_{k+m}\}^{[i]}, \{U\}^{[i]}, T, W, Y, Z \rangle\}$

2. Science as a Bertalanffyan system

a. The adduced definition of evolving systems barely resembles science. To make the similarities between science and systems in general more evident let us offer some platitudes about the notion of science we want to rely upon. We take science as a body of information that pretends to be true or at least to be justly acceptable. This stipulation excludes from science scientific publications (because they are only a means of communicating scientific ideas), scientific organisations, legal regulations for scientific activities, the activities themselves, financial aspects of research; the same holds true of research workers, teachers, instruments of research etc. Even the convictions of people elicited by science are left out from science itself. All these factors are outside science; they are included in the environment U of science.

b. Thus the relations $R_1, ..., R_k$ include only the relations that occur among pieces of information. Let us assume that the pieces of information are always given by sentences. The first and best known kind of relations which are under discussion are the numerous inference relations. Yet, they do not exhaust the set of relations that hold between pieces of information. Another important kind are the defining relations: *p* is defined in view of some *q*. There are also many classifying relations, relations which exclude some pieces of information on the basis of others. There are many kinds of similarity and dissimilarity relations, co-ordination relations of different types, etc.

c. The set of relations $\{R_{k+1}, ..., R_{k+m}\}$ includes all the semantic relations between *C* and the environment. *C* is now the body of knowledge and *U* is the world the knowledge is about. The relations just mentioned include many more elements. Research workers have been relegated into the environment. The same has been done with scientific instruments and the means of communication between scientists. Thus the relations for inventing a new concept, for inventing a new theory, the relations of using these and not those instruments are some of those that were mentioned at the start of the paragraph. The influence which scientific theories or hypotheses exert on people are relations that belong to the category under discussion. From our point of view, the most important relations included in the set now being discussed are all the applications of theories in our lives, all the technical innovations, new gadgets, new instruments, new machines, new medicines, new chemicals etc. Let us be aware that these new devices change the environment. What is also important, they exert some reciprocal impact on science. This is obvious in view of the new observational instruments and new convictions that become heuristic indicators for future research activities, which in turn give rise to new hypotheses and theories. All these relations may be generally spoken of as pragmatic relations between science and the world.

d. We have remarked that systems which develop must obtain energy and materials from their environment. We added that this cannot be done without discrimination. Precisely both of these necessities exist in the case of science and, more generally, in the case of theories. One cannot develop science without acquiring new information about the world. Even in deductive sciences, where all is given implicitly in the axioms and the rules of inference, one introduces new definitions and thus new notions in view of – usually – possible applications in the empirical realm. Although these new notions are only abbreviations of long and complicated aggregates of primitive notions, without definitions we could not manage to cope with problems. In all empirical sciences and theories the need to look for new data is obvious. Very similar considerations liken theories with systems in general when one takes into account the necessity to discriminate between the different factors inherent in science. In the case of a cell, its membrane filters out harmful material. In science or a theory, the methods used serve as ways of excluding from them any pernicious information. The simplest one is the language of the theory. Data that can in no way be expressed in the terminology of the theory (more exactly, cannot be constructed by means of the notions of the theory) must not be introduced into the theory. As an example, we are not allowed to include our delight at seeing stars during a dark night into an astronomical theory. Other methods of avoiding useless information are always connected with justification and there are thousands of ways to justify. They are the object of consideration of an extra science, namely methodology. These short remarks have been made to, tentatively, convince the reader that sciences and theories are systems in the sense of Bertalanffy. Further arguments to the same effect will follow.

3. Instrumental systems

a. In the preceding section science was spoken of as an entity that can incorporate information, can process information and can reject some parts of it.

In fact, science, as a set of information, does not exhibit any of these abilities. Sets of pieces of information do not act at all and therefore do not possess any ability to do anything and this is exactly true in the case of sciences and theories. Our way of speaking was metaphorical. All the activities mentioned are performed by human beings, primarily by research workers. People create sciences and theories, and use them in different ways to achieve their goals. Let us investigate the relationship of man and science a little closer.

b. Among systems one can easily discern special kinds of systems – instrumental systems.

Instruments are wholes that differ from their surroundings. As a rule, instruments are made up of parts, and specific relations hold among them. Each instrument exhibits different relations toward its environment. As a result, instruments have all the necessary features of systems in general. Instruments are in fact systems. This oversimplified description excludes, from the set of systems, simple tools, e.g. a stone we accidentally use to hit a nail that protrudes into our shoe and impedes our walking. This stone is, from the point of view of hitting nails, not composed of parts. All simple objects used by us will be called tools.

The most significant feature of instruments is that they are used by other systems. A thing is an instrument from the user's point of view, it is not an instrument according to its own virtue. An axe is an instrument in the view of the wood-cutter but not in the view of owls. What is more, an axe is an instrument in view of some goals, in view of, e.g. felling trees, but it is not an instrument with respect to shaving beards. The same thing can serve different aims in different moments. In such a case the thing can be split in two different instruments. A cannon, for instance, at one time is an instrument for killing while at some other time, when it is an object in a park and children play on it, the cannon is an instrument for play. After years, when the cannon has become rusted all over it ceased to be an instrument at all and was sent to a factory to be recast. So far we have pointed to the fact that the notion of instrument must be relative to at least three factors: the user, his goals and the time of using. Although there are more such factors, let us now ponder a little on the concept of aim.

When somebody tries to save some money he may save it to buy a car, to have a car, to use a car. The car itself is not the aim, but something that can be done or is done by the car or with the help of the car. The aim is then not a thing, but some events or processes or situations. Obviously, when somebody strives to achieve his aim it does not yet exist. When somebody saves money to buy a car, he has not bought it so far. Aims will be symbolised by the letter f (from the Latin *finis*). The aim which has been achieved will be symbolised by f^r.

Let us consider other factors that make things to be instruments. Once more let us consider two numerically different cannons that are otherwise identical. One shells the enemy every few seconds. The other was captured by the enemy but without any ammunition and serves as a very convenient object for the taking of snap shots. Both events, shooting and taking pictures, may happen at the same time. What makes one gun an instrument of killing and the other only an instrument of play? The circumstances of the use of the guns decide what kind of instruments they become. This example induces us to introduce the fourth factor. The circumstances which accompany the endeavours of the user may be quite different from the circumstances of achieving the goal. One saves money when one is relatively wealthy. When one has enough money to buy the car, the circumstances could change dramatically, maybe the money must be spent instead on a cancer operation for the saver. Such possible changes of fate must always be taken into consideration. The first kind of circumstances are those that exist at present, the other kind can be only contemplated or anticipated and may never be fulfilled. Because of this important difference the circumstances will be stratified. The set of the first circumstances will be symbolised by the letter *e* and the other, by the letter *a* (compare *anticipate*).

c. Everybody knows that one can crack nuts with the help of a hammer, but we also know that a nut-cracker better achieves the same goal. Are there ideal instruments with which we can accomplish everything we want and can do so unfailingly? Hardly so. In fact, most instruments help us to achieve our aims only with some probability. Some are better, some are worse. We have to distinguish between instruments on the grounds of grade of utility. This grade will be symbolised by the letter *n* which ranges over the set of real numbers. To simplify the matter, let us take into account only natural numbers.

We can write with a ball pen in different ways. The most convenient method to write with a ball pen is to hold the ball with our fingers of the right hand. A less convenient way is when we hold the ball pen with our teeth. The aims are, in both cases, the same. The user may be the same. The methods are different and they make the ball pen more (in first case) or less (in the second case) effective. To avoid the possibility that ball pens are instruments that exhibit both a high degree of utility and a low degree of utility we introduce a new factor: the method or way *g* of using the instrument.

d. So far we have introduced seven factors which make instruments out of things but we have considered only actual instruments. An axe which is kept for many years in the cellar without being used is not an instrument. Of course, one can easily define the notion of a potential instrument with the help of a Carnapian reduction sentence. The general scheme of the reduction sentence we have in mind here is the following: $[p \Rightarrow (q \Leftrightarrow r)]$. Let us consider what the antecedent of this implication is, this reduction sentence. We have already pointed to the fact that instruments are systems. The antecedent just notes this fact. Let S^I be an instrument. We must then explicitly state that S^I is a system. This statement is the antecedent. We have defined the general concept of a system and this definition is just the antecedent – with some minor changes. In particular, we must stipulate that in the environment *U* of *S* is something, let it be *V*, or rather somebody who uses *S*, and effects *U* with the help of *S*. The antecedent we are considering now includes a very complicated and very important concept: the concept of effecting or of acting upon something. This notion is so complex that we now abandon any analysis of it, otherwise we would be forced to abandon our problem and to ponder exclusively on the notion of action.

In the previous lines we left out the upper indicator. Now, in the full formula, it must be reintroduced, but instead of the complicated indicator $\{i \in I\}$ we shall use the indicator as already shown above, namely S^I.The upper indicator $\{i \in I\}$ is replaced by the simpler one but endowed with the same meaning, by I. Here is the antecedent. The quantifier for V will be attached implicitly only at the beginning of the whole definition

4. $S^I = \langle \{C\}^I, \{R_1, ..., R_k\}^I, \{R_{k+1}, ..., R_{k+m}\}^I, \{U\}^I, T, W, Y, Z \rangle \wedge \{V\}^I$ is a part of $\{U\}^I \wedge \{V\}^I$ affects $\{U\}^I$ with the help of $\{C\}^I$.

e. We already know that a system becomes an instrument when some factors are satisfied. All those factors must be cited in the definiendum. First of all, we must take into account both kinds of circumstances, these of kind *a* and those of kind *e*, as long as they happen to occur in the time space *I*. We must also include the aim *f* one tries to achieve, the methods *g* one uses (to be exact, *g* are the regularities that underlie his ways of proceeding) and lastly the degree *n* of aptness of the instrument. The definiendum is thus the following:

5. S^I in the environment $\{U\}^I$ is an instrument in degree n for $\{V\}^I$ in view of the aim f in circumstances $\{a\}^I$ and $\{e\}^I$ when used in the way g.

f. Now, we have to fix the most important part of the definition, i.e. the definiens. A system can be spoken of when it is used purposely and not handled by chance. The user must aim at the goal *f*. S/he must act according to the rules or regularities *g* and must obtain f^r with probability *n*.

6. $\{V\}^I$ aims at f and $\{V\}^I$ effects $\{U\}^I$ with S^I and $\{V\}^I$ obtains f^r and $\{V\}^I$ acts according to regularities g and the results f^r come into being in the circumstances $\{a\}^I$ and $\{e\}^I$ with probability n.

Now, there is no need to repeat (4), (5) and (6) in the full definition of the instrument. One has only to put them together in the general scheme $p \Rightarrow (q \Leftrightarrow r)$. (4) must be inserted instead of p, (5) instead of q and (6) instead of r.

g. Instruments can be used more or less successfully. They can be more or less accommodated to the goals the user strives to achieve. One fault is caused by the user. The second one depends on the structure of the instrument. The greatest problems arise when side-effects are considered. One may obtain his goal f, e.g. he has bought the car he dreamed of but, unfortunately, because of his unreasonable savings his wife decided to divorce him. In this obvious example there is no instrument but we can easily imagine other examples where instruments are used. Somebody tried to hit a nail with a heavy axe. She succeeded but accidentally she also hit her fingers with which she held the nail. The side effect was rather painful. A very frequent side-effect is fatigue or strain on the side of the user, sometime – when an instrument is used too often – occupational diseases may evolve, sometimes various kinds of mood affect his/her mind, sometimes instruments blunt his/her mental abilities or his/her moral sensitivities.

We shall try to define a good instrument now. We must take into account the undesired side effects. Some effects are undesirable because they are against the implicitly assumed aims of the user. The user tries, for example to remain healthy and then any illness caused by the use of the instrument is an effect which is unforeseen (usually) and is contrary to implicit aims, and/or contrary to the aim to remain healthy. Lets call the side effects $h_1, ..., h_l$. The numerous implicit aims may then be symbolised by $f_1, ..., f_j$. Now the utility of a good instrument evidently becomes very complicated. First of all, we must compare the importance of achieving the aim f with the importance of not obtaining at least some of the implicit aims which are precluded by the side effects. How can one compare the positive importance of earning one million dollars with the negative importance of losing one's health because of the work one performed to get the money? There must be a way to compare these otherwise incomparable dimensions. Secondly, each side effect hinders the achievement of one or more implicit aims with different probabilities. The utility of a good instrument cannot be identified with probability n as we have done in the case of the notion of an instrument. We must introduce a new instance which makes it possible to decide what is more important: the advantages brought by f^r or the disadvantages brought by the side effects. This instance is the superior aim F. Our example shows the collision between earning one million dollars and remaining healthy. One has *then* to choose: try to gain f or restrain from any action so as to avoid the undesirable side effects. The choice can be made only with a view to achieving a superior aim, e.g. being more happy. This superior goal may be called F. This kind of superior goal gives the subordinate aims f and $f_1, ..., f_j$ different but comparable values. The side effects hinder the implicit aims with different probabilities. Thus the ultimate utility of an instrument is rather difficult to establish. Because of the numerous factors involved and the different probabilities of the ultimate utility of a good instrument it is marked by the letter q although we are not able to calculate the numerical value of it.

h. Let us put together all our remarks made so far concerning good instruments and present something that resembles a definition.

7. S^I is for $\{V\}^I$ a good instrument with degree q in the environment $\{U\}^I$ in view of the explicitly accepted aim f, the superior aim F, the circumstances $\{a\}^I$ and $\{e\}^I$ and the rules g underlying the ways of acting if and only if

(1) S^I is for the user $\{V\}^I$ an instrument with degree n in the environment $\{U\}^I$ in view of the aim f, the circumstances $\{a\}^I$, $\{e\}^I$, the rules g, and $\{V\}^I$ achieves f^r

(2) there are implicit aims $f_1, ..., f_f$ of $\{V\}^I$ in the circumstances $\{a\}^I$ and $\{e\}^I$

(3) the user $\{V\}^I$ gets the effects $h^r_1, ..., h^r_l$ in circumstances $\{a\}^I$ and $\{e\}^I$ with the help of actions founded on g

(4) the achieved aim f^r contributes to obtaining the superior goal F of the user $\{V\}^I$ by degree u in circumstances $\{a\}^I$ and $\{e\}^I$

(5) the aims $f^r_1, ..., f^r_j$ contribute to the achievement of F by degree b in the mentioned circumstances

(6) the results $h^r_1, ..., h^r_l$ which differ from $f^r_1, ..., f^r_j$ hinder the achievement of F by degree d, and the degree u plus the degree b is greater than the degree d. The difference amounts to q.

4. Conditions of acceptance

a. From the very start we have assumed that science is a theoretical system consisting of pieces of knowledge. This treatment of science created some problems, the crucial difficulties being connected with the notion of a good instrument and thus with the notion of a good theory. All discussions about the truth of theories reduce to the question of distinguishing good theories from bad ones. First of all, we must thoroughly consider once again whether science is an instrument, because sometimes science is taken to have an autotelic value. It does not or should not serve anything but only itself. On the contrary, all other things should serve the aims of science. This view of science is closely connected with the conviction that truth is an autotelic value, superior to all other values.

b. When do we accept an ornament called inscription? Before we try to answer this question let us point to our assumption that an accepted inscription is the one we treat as satisfactory, as good. Thus accepted hypotheses or theories can be treated as good theories, at least for us and for the time being. Now, we will try to answer the question.

First of all, we must understand the inscription so that it becomes a sentence. We must know the simple inscriptions, rules of composition of these most simple elements of inscriptions, those that enable us to build short sequences (words) and those which enable us to put together short chains into much longer, with no definite limit, compounds of simple inscriptions (sentences). We also know the rules of reconstruction of different kinds: those that enable us to express our convictions in different ways as synonyms or those which enable us to make deductions. Inscriptions from the point of view of our definition of system are systems in spite of the fact that we cited only its parts and the relation between them. It is obvious that inscriptions happen to be written or read only in some circumstances and in some environments.

c. Let us consider the inscription *A horse is an animal.* We are inclined, without any hesitation, to accept this inscription as a true sentence and thus as a *good* sentence. But a person who knows only Urdu is not able to accept this simple inscription. Even after teaching this person the letters of the English language, the rules of word formation and the rules of grammar, s/he will not be able to understand the adduced inscription. So far we have not said what makes inscriptions understandable. In any event, we must take into account the people who are presented with inscriptions or sound sequences. The formula which predicts when an inscription can be accepted is, so far, the following: the person *v* accepts the inscription *x* composed from parts *x*.... according to rules *y*.... .

d. Let us take another example into consideration, namely the sentence: *The hot water is out of order.* The sentence can be accepted and rejected depending on what we have in mind. In the kitchen where the water is usually heated the sentence may be accepted. But in Iceland, in the vicinity of great jett fountains of hot water, this sentence is rather unacceptable, as no fountains cease to function. Obviously sentences are accepted depending on their ascription to some entities in the world or some entities in our imagined worlds, and how those entities happen to be in some way or another. The missing element needs to be filled in; we must take into account the so called reference. Let us give the formula which describes the conditions upon which we accept sentences, and leave out from it the simple parts and the rules of formation of longer inscriptions. After this simplification we get the formula: person *v* accepts the inscription *x* in view of the ascription to the reference *R*.

The next example is *The members of the royal family are visiting dignitaries* (Lakoff 1972, p. 545). This sentence is syntactically ambiguous. Therefore it can be accepted and rejected depending on the choice of its grammatical structure. As a result we must explicitly introduce the structure of sentences into our enumeration of the factors that underlie its acceptance. Our formula that collects these factors is now the following: Person *v* accepts the inscription *x* in view of the respective reference *R* and the structure *S*. Now, acceptance depends on three explicitly cited factors. One further example is *Tomorrow he'll write a letter*. This sentence can be accepted or rejected on the provision that we take into account two different days of utterance. Even more, we can assume that this sentence is simultaneously accepted and rejected. This happens when, e.g., we presume that different people have used this kind of inscription on different days. To avoid such difficulties we have to consider the circumstances of uttering or writing sentences. Montague's pragmatics is just the theory that pays attention to the problem of such dependencies. Our set of dependencies of acceptance must then include one member more and the formula which enumerates them becomes longer and points to four kinds of factors, the formula is now as follows: Person *v* accepts the inscription *x* in view of its reference *R*, its structure *S* and the circumstances *a*.

e. In a well known fairy-tale, we find the sentence: *Red Riding Hood loves her grandma*. Everyone who knows this fairy-tale is inclined to accept this sentence. But for a philosopher who exhibits his/her professional deviations it is doubtful whether to accept this sentence. The philosopher doubts whether we can accept a sentence that says something about practically nothing. Red Riding Hood does not exist, and will never exist. If Red Riding Hood does not exist, how can she love her granny. On the contrary, when the sentence is true, what is the reference of *Red Riding Hood* and of *The Granny of Red Riding Hood*? Is the reference in both cases the empty set, and are then Red Riding Hood and her Granny identical? Is the question at all applicable in the case of scientific theories which are of interest to us? Perhaps this question is important also in the case of theories. Theories resemble to some degree, and from some aspects, fairy-tales. The history of science tells us that once concepts of epicycles as real movements of planets were used. Such notions as phlogiston, the ether, the strange concept of the atom as something indivisible were used etc. We take these concepts to be void and empty and the statements about them resemble the statements of a fairy-tale. Of course from some points of view there are features that make us aware that fairy tales are rather different from even the worst theories. The problem is that the notions used today may later prove to be also empty concepts and the statements about them will acquire some similarity to the statements of a fairy tale. Today, we are not able to discern empty notions from not empty notions, we are not able to say which theory accepted so far is well founded and which is ill founded. Perhaps in the future some theories will be rejected and other more adequate ones accepted. In both cases, empty concepts and unknown non-empty concepts currently have to be endowed with the same kind of reference. Here I adopt the description of the reference of names and whole sentences put forward by J. Paśniczek (1998, *passim*). To put his conception into a nutshell and leave out his logic and semantics which enable us to handle both fictitious and non-fictitious sentences and their truths, let us say that a name's reference is a set of attributes (in fact a set of sets, as his system is extensional). If after some time of testing it turns out that the name's reference is non-fictional, then the number of independent attributes turns out to be infinite. This set is consistent in this special sense that from every pair of features $\{X, \overline{X}\}$ only one is included in the set of attributes, that make up the reference of a non-fictitious name. Additionally, the reference of the non-fictitious name is complete in the sense that from every mentioned pair one attribute is included into the set which determines the reference. Fictitious names refer to the sets of independent attributes, which are finite: only from some of the pairs mentioned above are attributes chosen. What is more, the reference may be inconsistent in this sense that both attributes from some pairs of the pointed pair can be included into the reference.

Now, let us return to *Red Riding Hood loves her Granny*. This sentence must be accepted on an analytical basis. The set of attributes that make up the reference of *Red Riding Hood* includes the feature loves her Granny. This last attribute is then included in the set of attributes that the reference of *Red Riding Hood* consists of. This sentence is false and must be rejected if we ask if there is an infinite number of independent attributes which are reasonably attributed to Red Riding Hood. This last question is rather empirical. It is rather obvious that Red Riding Hood includes an infinity of attributes in both cases: as a fictitious object and a non-fictitious one. But in the first case only some are ascribed to Red Riding Hood and the rest are only inferred. In the case of non-fictitious, real objects it possess an infinity of independent features that cannot be inferred from each other. At least in this way we treat objects of the macroscopic dimension. How many features can be displayed by an electron is not clear to me. At least it can exhibit many features independent of each other and dependent on its history, on the energy it has received, the atom it is attached or not attached etc. Presumably we do not know all the possible features of electrons, e.g. in extreme temperatures, and we are not able to infer from the accepted theories all the possible attributes of it. So far we do not have at our disposal a theory that explains all things and we have not the mathematical apparatus to deduce from the known theories all the possible details about the role of electrons. The main conclusion we want to draw from the preceding considerations is that from the finite set of attributes *R* we can draw many other features. As a rule one uses to this end not only deductive rules but also extra premises. Most of them are empirical in character. In the case of Red Riding Hood we can infer from the fairy tale and the children's empirical knowledge that Red Riding Hood often smiled to her granny, liked to talk with her, etc. The thus inferred attributes do not belong to *R*, they depend on the added premises, because of this we call them *E*. Children accept much unknown new information about Red Riding Hood only if the added new information conforms with their knowledge about people and with the general idea of the fairy tale incorporated in *R*. The possibility of accepting sentences that are not exclusively determined by *R*'s compels us to take *E* as an additional factor which influences the acceptance of sentences. We get the following formula which enumerates the factors:

> The inscription *x* is accepted by the person *v* in view of respective *R*, *E*, the structure *S* and the circumstances *a*.

f. When somebody lies purposely, s/he obviously does not accept the sentence s/he utters. It is also sure that his/her lie must be as similar to the truth or to accepted statements as possible. Only on the condition of this similarity can the liar be successful. From the lie one can infer new information on the basis of additional premises (and also without them, but this case is uninteresting now). Sometimes, rather often, the news will be fictitious and accepted along with all its consequences by naive people who do believe in lies, propaganda, advertising, etc. Thus, the lies endowed with the reference *R* determine additional outcomes *E*. Thus lies exhibit all the features of accepted statements but in spite of this the liar does not accept them. Why? There must exist an additional factor which hinders the liar from accepting his/her lies although other people might take them for granted. It seems that this omitted factor is our inclination to act upon accepted sentences. The liar is not ready to act according to his/her lie. Perhaps he only begins to act, to become more credible, or he simulates actions prompted by the lie. The liar knows that the lie does not result in achieving his/her aims when s/he acts according to them. S/he knows that s/he can achieve his/her aims when other people act according to the lie, but those people as a rule will not achieve their own goals. It is perhaps unnecessary to point that the link between the acceptance of sentences and acting is in most cases very indirect, sometimes there are, between the accepted sentence and the action, almost innumerable intermediate stages and factors. In any case, acceptance is connected with the inclination to act upon it in view of the aims the acceptor wants to achieve or accomplish. The formula which enumerates the factors of acceptance so far discovered is as follows:

> The inscription x is accepted by the person v in view of resp. R, E, the structure S, the circumstances a and the inclination to act in view of the aim f (of v).

g. When children listen to the fairy-tale called *Red Riding Hood* sometimes they wonder how was it possible that the wolf swallowed granny without biting her into pieces and wonder how was it possible that she was returned to the normal world without any harm. When confronted with such questions the story teller is in a rather miserable situation as s/he has to invent some mechanisms which could explain the unexplainable and has to do it in such a way that his/her listeners will understand and will be satisfied. The story teller cannot invent what s/he likes. S/he has to take into account the experiences the children have acquired in their lives so far. The explanations of the story teller are included in E but the choice of information and explanations are determined by the world (also cultural world) the children live in and know. Explanations given to European children may be, to some degree, different from the explanations offered to children in Brazil's interior. The choice of E depends on the world U and U must be treated as one of the factors which have some influence on the acceptance of sentences. Let us make the vicious remark that grown-ups do not differ greatly from children. Our formula is increased to:

> The inscription x is accepted by the person v in view of R and E, the structure S the circumstances a, the environment U and the inclination to act in view of the aim f.

h. Still some factors which determine acceptance are missing. We know already that acceptance is more or less connected with action. We know that we accept only some sentences, some are rejected and some neither accepted or rejected. Why is that so? Why is our behaviour apparently so strange? The usual answer to our question is very simple. We accept sentences which are justified, reject the sentences which are not justified and are neutral in the cases of sentences with neither positive or negative justification. This plain answer is true only when justification is understood in a very broad sense. Children accept the sentence to the effect that the wolf swallowed grandma by virtue of the authority of the story teller, e.g. their mother. We accept sentences mostly on the authority of books, journals, TV news, teacher's statements or on the authority of more experienced friends. There are thousands of ways to justify sentences, some of them are respectable, others valueless and still others are simply criminal. We won't go into the problem of the value of different methods of justification, here let us only state that acceptance depends on justification which will symbolised by *g*. Now the formula which exposes the factors of acceptance is as follows:

> The inscription *x* is accepted by the person *v* in view of *R* and *E*, the structure *S*, the circumstances *a*, the environment *U*, the inclination to act in view of the aim *f* and the methods of justification *g*.

i. We have already mentioned that there are numerous different methods. Sometimes the methods *g* are employed without any reflection and without any possibility of making a choice between different methods. However, the method employed is rather often dependent on our decision and on the circumstances which induce our decisions. The circumstances just mentioned are different from the circumstances *a* of uttering the sentence. As a rule, the circumstances which affect the methods of justification precede circumstances *a*. It is worth citing in our formula these new kind of circumstances because they exert influence on the methods and the methods determine acceptation to a large extent. The new circumstances will be called *e*. The next formula is enlarged by *e*:

> The inscription *x* is accepted by the person *v* in view of *R* and *E*, the structure *S*, the circumstances *a*, *e*, the environment *U*, the inclination to act in view of the aim *f*, and the methods of justification *g*.

j. Often we have to do with hypothesis *ad hoc*. Generally, these hypotheses are treated as valueless. They are introduced into discourse only because we hope they will become normal, informative hypotheses, or they are already such hypotheses although for the time being this fact is not known. When somebody proves that a hypothesis is only an *ad hoc* means to save another theoretical conception this hypotheses is left out. What makes hypothesis *ad hoc* so valueless?

They are introduced to explain away one specific difficulty or a very limited number of them. As a rule we do not link them with any expectations of problem solving which are not known so far. They simply are not conceived of to extend our knowledge, i.e. the aim they have to serve is confined. A non-*ad hoc* hypothesis must help us to solve many problems. It has simply, many goals to obtain. Then the acceptance of a hypothesis is dependent on the aims that supplement the aim f. The additional aims which contribute to the acceptance of a sentence may be symbolised by $f_1, ..., f_j$. All aims may be only intellectual. It is not necessary that our aims are physical or are applications of the accepted sentence in industry, etc. Sometimes the only immediate aim is simply a change of ones convictions, ones attitudes, emotions etc. Our formula of acceptance is as follows:

> The inscription x is accepted by the person v in view R and E, the structure S, the circumstances a, e, the environment U, the inclination to act in view of the aim f, the methods of justification g and the additional aims $f_1, ..., f_j$.

k. We have found that acceptance is connected with actions. The first actions consist of the justification proceedings. They can be very complicated, as in the case of building and starting a super collider, or very simple, as in the case of opening our eyes to see somebody's hair-do or waking up to listen to the latest news on the radio. The second kind of activities are connected with one's endeavours to achieve his aims on the basis of knowledge formulated in the accepted sentence. Both activities contribute to the acceptance of the sentence. The first kind of activities are obviously a necessary condition of acceptance. But the others are also essential for sustaining acceptance. Let us imagine that somebody accepted the sentence that eating carrots makes it possible to get rid of a cataract. S/he eats (let us call this activity *d*) two pounds of carrots every day but finds after a year that his/her cataract is worse than before. Perhaps s/he will stop accepting the sentence on behalf of it s/he tortured him/herself with eating large amounts of carrots every day. Unfortunately, both kinds of activities bring about some unforeseen or undesired effects. The construction of a super collider prompts some protests in parliament because of the supposed increase in taxes. This kind of side effect exerts no influence on the acceptance of the sentence that this gigantic accelerator can give some clues as to the existence of Higgs particle. But one can imagine that the results of experiments performed in the super collider will be inconsistent with some accepted theories, e.g. the electro-weak theory in physics. The second kind of activities, which we call here up to now the application of accepted theories, are the almost infinite sequence of adoptions of the theory in successive new circumstances. These applications may result in numerous quite unexpected incidents. These unforeseen events are proof that the applied theory had some gaps or was wrong in some details. When the theory of the combustion engine was established nobody gave any attention to the composition of the combustion gases with all their cancer producing substances. Thus, the theory was not complete. Only now, after many damaging experiences have we supplemented the theory. The most damaging feature of incomplete theories is that they are treated as complete ones. We can not imagine in advance that there is a gap in the theory somewhere.

Both kinds of side effects, together with the expected ones, let us call them $b_1, \dots b_k$. Some of them are known from the very beginning, i.e. from the time we have accepted the sentence. Some of them become evident much later (after having performed *d*). Those which are known from the beginning are either identical with the expected f^r and some of $f^r_1, \dots, f^r_j$, some, let us call them $h^r_1, \dots, h^r_m$, in different ways hinder the achievement of some of $f^r_1, \dots, f^r_j$. This happens very often. One arranges an experiment in which the desired aim has been obtained in the form of confirming this or some other way that our hypothesis, whilst providing some result which abolish some other theories. As a rule, one must consider the question of what has been done wrong or which result is more important: the one in favour of the new hypothesis which abolishes the old, or the ones that are in favour of the old theories and which are incompatible with the new one. Discussions to this effect can be very fervent and long and criteria of value are referred to. Discussions between the followers of Ptolomeo on the one side and Copernicus and Kepler on the other side give a good example of the appeals of criteria of value.

To decide whether acceptance of the given case is advisable one compares the value of the achieved f^r_i and the negative value of $h^r_1, \dots, h^r_m$. We assume that the aims which we strive for have a positive value. All the events we try to avoid have a negative value for us. Thus, to compare two values is to find out which aims connected with the two values conform better with our supreme goal or goals *F*. This very short remark does not go into the very important problem concerning the choice of aims, relativism of aims etc. These problems are entire worlds of questions and difficulties and we cannot go into them here, simply because of lack of space and the desire to reach the end of these considerations, so we will only introduce into our formula of acceptance the additional factors *F* and $h^r_1, \dots, h^r_m$, arriving at:

The inscription x is accepted by person v in view of R, E, the structure S, the circumstances a, e, the environment U, the inclination to act in view of the aim f, the additional aims $f_1, ..., f_j$, the methods of justification g, the activities of application d, the side effects $h'_1, ..., h'_m$ and the superior aim F which enables us to compare the values of the f^r's with the values of the h^r's.

5. Science as an instrument

a. We already know what an instrument is, and what a good instrument is. In the last section we pondered a little on the conditions of acceptance of sentences, i.e. on the conditions of sentences that are good for us. Sentences have been treated as substitutes of hypothesis and whole theories. We have considered, then, the conditions of good theories. Now, we can examine whether a good theory is a good instrument and whether a theory is an instrument at all. First of all let us compare the factors that decide that something is a good instrument and the factors that decide that we accept a theory and in this way we treat it as a good theory. If the factors coincide one can suppose that the theories are instruments, and that the accepted theories are good theories.

b. Instruments are systems. The same holds true in the case of sentences and even theories. They consist of parts interrelated by S structures. These rules are both rules of construction and rules of transformations, peculiarly rules of inference. The S structures, although only slightly touched upon, are counterparts of the relations $R_1, ..., R_k$ shown in our characteristics of the system The reference $\boldsymbol{R}$ (to make discrimination between the schematic letters which belong to the definition of system and instrumental system and the letters of the formula concerning acceptance, the last letters will be given in bold) and its consequences $\boldsymbol{E}$ are parts of $\boldsymbol{U}$. The relations between the sentence constructions and the reference $\boldsymbol{R}$ and $\boldsymbol{E}$ are left out as obvious. They are only hinted at by the word *respective* preceding $\boldsymbol{R}$. The semantic rules that connect sentences and their parts with references are counterparts of the relations $R_{k+1}, ..., R_{k+m}$. $\boldsymbol{R}$ and $\boldsymbol{E}$ are included in $\boldsymbol{U}$. Thus U consists of sets of attributes; some of the sets are fictitious and some real. Of course, theories are entities that last some time and during this time they change. Therefore the $\boldsymbol{R}$, $\boldsymbol{E}$, $\boldsymbol{U}$ and the letters which will be spoken of later must be confined to a period of time that has been taken into account. The references etc. are, therefore, designated $\boldsymbol{R}^{[I]}$, $\boldsymbol{E}^{[I]}$, $\boldsymbol{U}^{[I]}$. Evidentially, the person v mentioned in our formula of acceptance acts at the same time I and the person is the user of the sentence, i.e. the theory. This person understands the theory, s/he tries to apply it technically or at least psychologically by way of changing his/her own convictions and attitudes, or the convictions and attitudes of other people or even the attitudes of animals. The user, now described as $V^{\{I\}}$ does not understand the sentence when s/he is not interested in getting any help from the orientation of $U^{\{I\}}$. Such strange behaviour can be observed in cases of pupils who hate to learn. One can explain to them, hundreds of times, the most simple things and they still will not understand although they could be much better at handling a computer, for example, than the teacher, for their brains are in perfect order. Having an aim $f^{\{I\}}$ is therefore a necessary condition for understanding, and the more for accepting sentences. Sentences and theories behave exactly as instruments. As a rule a theory has many tasks to perform. It is often underlined that theories summarise the hitherto scattered knowledge; that they explain facts or less general theories; that they make up a coherent view of a domain; that they become applicable in practice; that they enlighten our minds or that they confuse our minds; that they induce us to act, etc. etc. These are the aims $\{f_1, ..., f_j\}^{\{I\}}$. Only when a hypothesis performs many of these aims it is not a hypothesis *ad hoc*.

We have discriminated two kinds of activities connected with the acceptance of hypotheses. One kind is related to the justification that was set before the acceptance of the hypothesis. The other is concerned with putting the hypothesis into practice (activities *d*). If our endeavours to achieve our aims, on the basis of a hypothesis, are as a rule ineffective we begin to doubt whether the hypothesis has been accepted on a sound basis. Later we simply try to find a new and more effective hypothesis and the old one is simply rejected. Exactly the same thing happens with other instruments. We construct them, try them on a small scale and then use them on full scale. Sometimes the instrument is useless for full scale applications. The ways *g* of justification are the counterparts of the construction of an instrument. In the case of theories and in the case of other instruments we have to use activities of type ***d***. The difference lies only in the physical or intellectual characteristics of these activities in the case of other instruments and in the case of theories. The mentioned inefficacy of applications of theories is sometimes caused by a simple inability to obtain the aims, sometimes this is because the side effects of our activities are impeding the additional aims $f_1, ..., f_j$. These aims vary from well defined events up to very vague ones, such as: to become happy or rich or satisfied in as many ways as possible. The activities ***d***, based on the accepted theory, as a rule enable us to achieve the main aim f, otherwise the theory would not have been accepted. However, successive failures to obtain the other, implicit aims, in spite of repeated activities ***d*** based on the accepted theory in question, can result in doubts concerning the theory after some time. This important conclusion refers to theories conceived of in the most general way, general views of life and world included. In the case of such a change of attitude towards the theory, the value of having gained the aim *f* is tempered by the discouraging negative values of the failures, and then the superior aims and values ***F*** are taken into account. In fact, this holds true of both instruments and theories in general. The disappointments with science that arise every 10 years are a symptom of the mechanism described. The implicit aims people set with science simply are not put into effect. This simple remark makes us aware of further problems.

c. The acceptance of theories is closely related to the aims of the users of the theories. This factor has been denied for centuries. When acceptance was dependent on somebody's aims s/he was accused of wishful thinking, almost the worst sin of a scientist. One can be accused of this error only when one accepts a theory before its justification, i.e. before having performed activities of the kind ***g***, or when one accepts a theory in spite of a negative justification. However, we will never be able to fully discern wishful thinking from sound justification. The problem arises from the necessity to compare the achieved and desired aim *f* with undesired side effects obtained during the justification. But we cannot accuse anybody who recalls his/her acceptance when his/her implicit aims $f_1, ..., f_j$ cannot be achieved, even after long periods of time and great endeavours when performing actions ***d***, of wishful thinking. Those who do not revoke their acceptance of the theories in spite of their uselessness in the long run are simply dogmatic.

d. The implicit aims also give rise to further remarks. The first of them is to the effect that we do not known them consciously but they nonetheless exert some influence on our acceptance of theories. Those implicit aims may be indirectly linked with the accepted theories. The theories of natural science are rather loosely connected with such aims as a safe life, a hope that life will become happier, etc. The disappointment concerning these aspects of life, in spite of the development of science, prompts people to underestimate the value of scientific theories, and this trend is increasing. The reasoning behind this attitude is the following: such a mighty instrument as science, which determines almost everything, does not make the quality of life better; the quality becomes rather worse and, perhaps, it is endangered in a most fundamental way; thus, something must be wrong with science itself and not only with its application. The Socratian way of thinking is quite clear about this kind of reasoning: when something goes wrong then the theories that were accepted and used, peculiarly the theories which were treated as the most outstanding ones, must be wrong and therefore rejected. Very often the failure of achieving the implicit aims manifests itself only years, or even tens of years after the acceptance of the respective theories. Some pedagogical theories that have prevailed for many years and enabled us to have the general motivation to live and co-operate with other people and help them warm-heartedly etc. only now, after several generations, reveal their true value. Another striking example of change over time, after which we encounter the unpleasant side- effects of a theory that prompts us to reject it in its old form, are the theories concerning the chemical reactions in combustion engines. At the end of the 19th century nobody cared about the side products of combustion. Now, its by-products that give rise to cancer are studied very intensely. Those old theories were not complete but were treated as far more adequate. Then partial truths were achieved, but as we know, the worst lies and errors are half-truths.

e. So far there is a wide gap between the scientific justifications of theories and the popular evaluations of theories after some tens of years of using them. Scientific justification concentrates on the achievement of the principal aim f. Only a few of the implicit aims $f_1, ..., f_j$ are taken into account in scientific research procedures. To the contrary, the popular vote concerning the acceptance of theories is much more closely linked with numerous implicit aims. Therefore, there are discrepancies between the acceptance of theories by research workers and average people although they are exposed to massive information campaigns and school education to teach them to accept science without doubt. All the complaints of people concerning the technical applications of science are not treated as instances of falsifications of theories. The complaints are not treated as some indications that the present theories are to be amended and supplemented. They are treated instead as indications of the political misuse of science. The Socratian attitude: bad acting is always connected with some errors of the accepted theories, is simply unknown or is rejected. Nobody asks why missuses can happen in spite of the generally accepted notion that people do not act without thinking, and that this thinking is to a great degree determined by convictions acquired in the process of education and not only from interests. Widespread social satisfaction or dissatisfaction must be symptoms that both the applications of theories and also the theories themselves are not entirely correct. The only problem, and the most important one, is that we accept simultaneously thousands of theories and it is very difficult to find out which of them are the culprits of our mental discomfort. When complaints are widespread then it is natural to look for the wrong ones not in very specialised theories but in very general and widely known ones. Perhaps these theories have no definite formulation and because of that are rather philosophical in character.

f. Usually aims are chosen freely. As acceptance is dependent on these aims different people may accept even incompatible theories. Acceptance is then left to one's unrestricted free choice. At least things look like that on the surface. In fact, the aims are chosen so as to conform to superior goals. But, in turn, the superior aims are seemingly also very subjective and differ from case to case, from man to man. Actually, in many cases this is what happens, but when we look for further superior goals we find one that is almost always assumed. This aim is simply the existence of the user of the theory (in the next section we shall ponder a little on the problem of who or what is the user of science and the theories). The very simple fact is that without taking account of this aim the user is, sooner or later, doomed to perish. Maybe this statement seems to be too pessimistic but, in fact, it is very realistic. Somewhere this most superior goal is always assumed, at least implicitly. Scientific theories, philosophical theories etc. do not strive for the death of the users of these theories (rather they strive sometimes for the death of the opponents of these theories). Most often the goal of the users of theories is, directly or indirectly, their existence and life and development. Thus the ultimate goals of the users (their culture, religion, nation, even environment) are not included in the domain where the proverb: *de gustibus non est disputandum*, is being applied.

g. Who is the user of theories? Are the users of theories, especially of scientific theories, those who decide about the research work? Are the users of scientific theories those who give financial support for the research work? In the first case the users of scientific theories are research workers. In the second case the users are states and financial empires. There is still another possibility. The users of scientific theories are those people who get the advantages from scientific theories and along with them also the disadvantages. In fact, this is the whole of mankind. Even more, the results of the application of theories, the good and the bad ones, affect the whole globe with all its creatures. Perhaps the results of the applications of theories will also have some influence on the surrounding cosmos. If true, the user (the scientists the whole society or the whole ecosphere or even the cosmosphere) determines goals which are very different from our everyday aims. The existence and development of the user (conceived of in the last way) may prove to be very different from our current social well-being.

h. The last sections of this talk have not solved any major problem but they have shown that in the conceptual frame of instrumental systems one can deliberate almost all the problems that are now under discussion in different domains, with the help of different and incomparable conceptual schemes. In the conceptual frame of instrumental systems one can consider the problem of the objectivity of theories, of their dependence on social tendencies; the problem of the political aspects of science, e.g. the problem of financial subordination of scientific research, the problem of the goals of science and its ecological implications. In this framework one can compare, systematically, theories and whole sciences with such structures as culture, philosophy, myth, religion and even political conceptions. Some hints to this effect have been made in this paper.

Department of Methodology
Philosophical Faculty
Catholic University of Lublin
Al. Racławickie 14
20-031 Lublin
Poland

REFERENCES

Lakoff, G. (1972). Linguistic and Natural Logic. In: D. Davidson, G. Harman (Eds.), *Semantics of Natural Logic*. Dordrecht: Reidel, pp. 545–665.

Paśniczek, J. (1998). *Logic of Intentional Objects. A Meinongian Version of Classical Logic*. Dordrecht–Boston–London: Kluwer.

Poznań Studies in the Philosophy of the Sciences and the Humanities
2000, Vol. 72, pp. 61-88

Adam Grobler

EXPLANATION AND EPISTEMIC VIRTUE

ABSTRACT. It is argued, contrary to Van Fraassen's theory of the pragmatics of explanation, that explanatory power is an epistemic virtue. Lipton's best-explanation account as a solution of the problem of induction is discussed and amended by providing a formal criterion of relative "loveliness" of competing explanatory hypotheses. The criterion on offer can also be applied to the question of rationality of a possible revision of background knowledge (or a paradigm, or scientific research programme, etc.) and hence provides a solution to incommensurability problems. Next, the Inference to the Best Explanation Principle is shown to operate, apart from the level of scientific theories, at the methodological and epistemological levels as well. In conclusion, explanatory power considerations are claimed to be a guide to the truth, realistically conceived.

1. A preliminary explanation of "explanation"

Before we consider the question of the epistemic value of explanatory power, i.e. of the relevance of explanatory power to the idea of truth, we have to explain the concept of explanation itself. Some philosophers, e.g. Koertge, are ready to abandon the idea that explanation is one of the functions or aims of science, on the grounds that there is no viable account of what a scientific explanation actually is.

Hempel, summarizing his well-known classical account (1965), gives two criteria for what is an explanation:

> *explanatory relevance:* "the explanatory information adduced affords good grounds for believing that the phenomenon did, or does, indeed occur."
>
> *testability*: "the statements constituting a scientific explanation must be capable of empirical test."

One reason of the inadequacy of the above criteria is the presence of well-known paradoxes of the asymmetry of explanation. The most famous example is the flagpole. Assuming the theory of optics, and measuring the angle of the elevation of the sun, we can deduce, using trigonometry, the length of a flagpole's shadow from the height of the flagpole as well as its height from the length of the shadow. Hence, according to Hempel's criteria, the height of a flagpole explains the length of its shadow no more than the length of the flagpole's shadow explains the height of the flagpole, which is counter-intuitive. Another example is the barometer. On the hypothesis that a barometer reading drops if and only if a storm is coming, one can deduce the drop in a barometer reading from the coming of a storm or the other way round. Yet it is counter-intuitive to say that the drop in a barometer reading explains the coming of a storm as much as the coming of a storm explains the drop in a barometer reading.

The paradoxes of asymmetry are removed in the causal theory of explanation, most notably proposed by Salmon (1984). It is the coming of the storm (according to the relevant scientific theory) which causes the falling of the barometer reading, and therefore the former explains the latter rather than the other way round. Similarly, it is the flagpole which causes (casts) the shadow, and therefore the former's height explains the latter's length.

The causal theory of explanation, however, involves the notoriously obscure concept of cause which is traditionally metaphysically-laden. Moreover, accepting the causal theory commits one to the requirement of reducing all explanations to causal explanations. This seems to be too restrictive and hardly a methodologically fertile requirement. Consider intentional explanations that play a significant role in psychology or the study of crime detection. Even if we confine ourselves to natural science, to the exclusion of behavioural sciences, one can doubt whether we are entitled to rule out the possibility of non-causal regularities. It is difficult to imagine at present any alternative to causal explanations in natural science but, if one did appear, it could offer new prospects for solving the conceptual difficulties of quantum physics.

Instead of appealing to causality, I suggest to require from an explanation that there be a one-many relation between a relevant part of the *explanans* and the *explanandum*. A flagpole has just one height but many shadows of different length, according to different positions of the sun. Similarly, for one storm there are (possibly) many barometer readings that will have dropped on the occasion of its coming. The reverse of a one-many relation is a many-one relation. This is sufficient to block the paradoxes of asymmetry without absolutization of causality.

One may object saying that the one-many relation between phenomena is just a symptom of a causal dependency (and of the sense of this dependency). Thus my proposal to avoid invoking causes in explanations is spurious. In response, I can point to one-many relations between a part of the *explanans* and the *explanandum* in explanations of a non-causal type, e.g. intentional or functional explanations. Hence, the one-many principle admits the possibility of explanatory factors other than causal ones, including those of as yet unknown types.

The one-many principle, in addition, rules out superfluous "explanations" which are permitted in Hempel's deductive-nomological model. Consider the following example:

(L) The melting point of copper is 1063°C.
(p) This piece of metal is of copper.

(q) This piece of metal melts at 1063°C.

This argument fullfills formal Hempel's criteria for being an explanation. In my proposal, such Molierean "explanations" are excluded, for the relation between p and q is one-to-one (and thereby does not block the symmetry) rather than one-many.

The one-many principle also permits to distinguish explanatory laws from phenomenological laws, i.e. the laws which themselves call for explanation. Consider, for example, the Boyle-Mariotte law and the following exemplification of Hempelian schema:

(L) $pV = \text{const.}$
(p) The pressure of this sample of gas changed from p_1 to p_2.

(q) The volume of this sample of gas changed from V_1 to V_2.

L determines a one-to-one relation between changes of pressure and changes of volume of the sample of gas in question. Consequently, p and q are interchangeable and the schema does not represent an explanation (The Boyle-Mariotte law is not explanatory).

Other difficulties concern statistical explanation. One widely discussed example is that of syphilis and paresis. Some people suffer from paresis if they have suffered before from syphilis but nobody suffers paresis if they have not suffered from syphilis. Someone's suffering from paresis cannot be explained, however on Hempel's account, by their previous suffering from syphilis, for the probability of contracting paresis is low.

For this reason Salmon (1971) proposed to replace Hempel's requirement of high probability with the requirement of statistical relevance. By definition, A is positively (negatively) statistically relevant to

E if and only if $\mathbf{P}(E/A) > \mathbf{P}(E)$ (resp. $\mathbf{P}(E/A) < \mathbf{P}(E)$). In this account, however, one cannot distinguish explanatory coincidences from coincidences which themselves call for an explanation. Or, in a causal theorist's parlance, one cannot tell sheer statistical relevance from causal relevance. To improve upon this, Salmon offered in his (1984) a statistical-causal theory of explanation. This, however, as it was said before, is not my favourite choice.

Other difficulties of the statistical relevance approach have been pointed to by Cartwright (1983). In her example, one cannot explain the survival of a plant by its being sprayed with defoliant of .9 effectiveness. The difficulty is removed by van Fraassen's principle of correct contrast. In his (1980, p. 111) he states: "We can explain why John, *rather than* his brothers, contracted paresis, for he had syphilis; but not why *he*, among all those syphilitics, got paresis" (my emphasis). Similarly, we cannot explain why this particular plant survived; but we can explain why that particular plant died, *rather than* another one, that was not sprayed with defoliant.

Now, we can see that the one-many principle and the principle of correct contrast help to explain the very concept of explanation. Before we sharpen this preliminary explanation of "explanation", let us consider the question of whether the explanatory power of a scientific theory counts for its epistemic virtue.

2. Van Fraassen on dispensability of explanation

According to van Fraassen's constructive empiricism, explanatory power is not an epistemic virtue at all:

> that the observable phenomena exhibit these regularities, because of which they fit the theory, is merely a brute fact, and may or may not have an explanation in terms of unobservable facts "behind the phenomena" – it really does not matter to the goodness of the theory, nor to our understanding of the world (van Fraassen 1980, p. 24).

Let T be a theory accepted because it explains another successful theory T'. Now: "if the mere fact of postulating regularities, without explanation, makes T' a poor theory, T will do not better" (*ibid*, p. 25). In other words, if *the only* virtue of a theory is to provide an explanation for another, otherwise successful theory, it is not a virtue at all. This is perfectly correct: such a theory T would be *ad hoc* relative to T' and would add nothing to its descriptive power. But suppose that there was one more theory T'' such that: (i) the union of T' and T'' is not a unified theory in Watkins's (1984) sense, that is, the union of T' and T'' has no testable consequences apart from those of T' and of T'' taken separately and: (ii) T explains T' and T explains T''. In such a case T is not *ad hoc* relative to both T' and T'', and goes beyond the descriptive power of either theory. As we shall see, van Fraassen's dispensability arguments apply to *ad hoc* explanations only.

Let us have a look at the historical examples van Fraassen quotes to support his dispensability thesis. The chief example is Newton's saying "*hypotheses non fingo*": his refusal to give an explanation of gravity (van Fraassen 1980, p. 94). In truth, however, Newton did not refuse to put forth hypotheses at all – what is, if not a hypothesis, his principle of action at a distance? – nor did he refuse to give explanations: his theory explains Galileo's and Kepler's laws. Newtonian explanation of Galileo's and Kepler's laws is exactly one of the *non-ad hoc* type described in the previous paragraph. In contrast, it was simply inconceivable to give a *non-ad hoc* explanation of gravity in Newton's time. On the other hand, it was the striving to explain gravity that resulted in the formulation of Einstein's general relativity theory. The latter, again, stands to classical gravity and special relativity theories in the relation described in the previous paragraph.

Similar to Newton's attitude towards gravity is the reluctance of today's quantum theorists to give an explanation of the correlations considered in the famous paper by Einstein, Podolski and Rosen. Van Fraassen rightly points out that such "metaphysical extensions of the theory (if indeed possible) would be philosophical playthings only" (1980, p. 95). This, again, is a case for the dispensability of *ad hoc* explanations rather than explanations in general. If EPR correlations share the lot of the previously mysterious gravity and finally find a *non-ad hoc* explanation, the scientist will only be too happy.

Another van Fraassen argument is borrowed from Kuhn's observations of explanatory "losses" in the history of science. A favourite Kuhn example – which is utilized by van Fraassen (1980, p. 111) – is Lavoisier's giving up, once he accepted the oxygen theory, the "phlogistic" explanation of the metallic lustre of metals. What is worth noting here is that Lavoisier was rewarded for this with explanations hitherto unavailable. Thus his alleged "loss" was, at worst, an "exchange". And this exchange was by all means profitable, taking into account that "the lost" explanation was, in fact, "put in pawn" rather than "spent". What I mean is that Kuhn's "losses" can be regained just like something pawned. There was nothing in the oxygen theory which precluded a future explanation of the metallic lustre of metals within a more developed version of the theory or within some other theory which would be compatible with the oxygen theory. A similar pattern can be observed in connection with the replacement of Cartesian physics with Newtonian physics. There were "losses" of the vortex theory's explanation of the coplanarity of the planet's orbits, which was regained by the theory of the evolution of the solar system added to Newton's theory. Hence Kuhnian stories by no means support the thesis about the scientist's neglect of explanation. To the contrary, they illustrate the scientist's striving for new explanations, for which they are willing to pawn their old explanations hoping that eventually they can regain them and thereby improve upon their epistemic balance.

In summary, historical examples of the scientist's alleged rejections of the demand of explanation are, in fact, either rejections of *ad hoc* explanations or putting some explanations "in pawn" in order to gain some other, hitherto unavailable, explanations. They do not support van Fraassen's dispensability of explanation thesis.

This reinterpretation of historical evidence in disfavour of the dispensability thesis is not, in itself, the evidence to its contrary. After all, the great scientists of the past may have been mistaken on the methodological point concerning explanation just as they have, more often than not, been mistaken in their theories. Van Fraassen may well be right in claiming that Darwin's dictum, which he quotes in his (1980, p. 98): "It can hardly be supposed that a false theory would explain, in so satisfactory a manner as does the theory of natural selection, the several large classes of facts above specified" suggests but that "the explanation-relation is visible before we believe that the theory is true" (*ibid*), which does not support Harman's (1965) principle of "inference to the best explanation". What Darwin had in mind in the quotation was the relation between his theory and *established* facts, which makes his explanation an *ad hoc* one. For this reason Popper regarded the theory of evolution as a metaphysical programme for science rather than a scientific theory proper. Again, van Fraassen's argument applies to *ad hoc* explanations and not to explanation in general.

Still, van Fraassen maintains that explanatory power cannot be relevant to truth, since there is a categorial difference between the two. The former is a pragmatic virtue while the latter is a semantic property. "To praise a theory for its great explanatory power . . . is not tantamount to attributing to it special features which make it more likely to be true, or empirically adequate" (van Fraassen 1980, p. 89). On van Fraassen's constructive empiricism, the only epistemic value is empirical adequacy. In this view, the epistemic value of a theory T which explains theories T' and T'', if any, is not that the former explains the latters but, rather, that the former is empirically adequate to a higher degree (if it is) than the latters. In such cases, "explanation-relation" is, as it were, a sort of epiphenomenon. "But it might be arguable that, *for purely pragmatic* (that is, person and context related) *reasons*, the pursuit of explanatory power is the best means to serve the central aims of science" (*ibid*).

It is suggested, then, that a scientist's striving for explanation is of purely instrumental value for the only possible epistemic gains: the increase of empirical adequacy of accepted theories. This results in working out "a picture" which "our language is guided by" (*ibid*, p. 14), a picture which is not to be believed. Explanatory theories which generate research programmes – "pictures" – are the instruments of knowledge rather than its parts. Their explanatory power is just the suggestiveness of the picture. In this picture, it remains unexplained why the explanatory power of the theories underlying research programmes promotes their heuristic power so as to suggest theories of greater empirical adequacy. Newton-Smith, an opponent of constructive empiricism though he is, concedes in his (1989) that one can claim the positive correlation between explanatory power and empirical adequacy to be an inductively established bare empirical fact. Yet even van Fraassen, when he says that something is "merely a brute fact", has in mind a regularity which, though unexplained, is nevertheless envisaged by, or incorporated into a theory. And the issue is what are the theory's epistemic virtues. In the next section I shall point to the role of explanatory considerations in inductive reasoning, the role which transcends pragmatic dimension.

3. Explanation and prediction

The main point I want to make is that the explanatory power of atheory not only adds to its predictive power but also does this in a way which lends it the kind of support that is not available from the mere satisfaction of its predictions. The discussion of Sec. 1 can, in fact, be described as an attempt to separate the concept of explanation from that of prediction. The shortcomings of Hempel's account, or even Salmon's S-R model, consist in imposing on explanation the very pattern of scientifically valid prediction. In contrast, the amendments offered in Sec. 1 are intended to account for those features which make explanation something more than mere prediction (or retrodiction). Let me illustrate the difference with an example.

Suppose we have tossed a penny 100 times, and tails have come up every time. The probability of this result, on the assumption that the coin is regular, is extremely low. On the received view on induction, this makes us suspect, or even believe, that the coin is biased. We can put forth a suitable hypothesis, make predictions on its grounds, and test whether the predictions are correct. My claim is that even if the predictions appear correct, the epistemic value of the hypothesis arrived at in this manner is not very high, in the absence of any explanation. The argument I am offering below concerns, in fact, the old Hume's problem of induction in the modern, Bayesian conceptual setting.

The only reason for accepting the hypothesis of a bias is that, on the assumption that the coin is regular, the probability of having 100 tails out of 100 tosses is low, much lower than that of the same result, calculated in accord with the hypothesis in question. On the other hand, if the coin is regular, the result of 100 tails of 100 tosses is not any less probable than any other distribution of heads and tails over the series of tosses. Say, 50 tails on every odd toss, and 50 heads on every even one. Or 50 tails in the first 50 tosses, and 50 heads in the remaining ones. (This is the paradox that is well-known to bridge-players: the probability of receiving 13 spades is equal to that of receiving any other hand. Yet any player would find it remarkable if s/he one day received 13 spades. It is only the player's disregard of the difference between small cards which creates the sense of paradox. In the case of the coin, we tend to disregard the ordering of heads and tails in favour of their number.) Consider the last case. What are we to believe in the middle of such an experiment, and what in the end? If we are to change our mind as the experiment goes on, why are we to stop after 100 tails and not try another 100 tosses with the possible result of 100 heads? These questions cannot be answered in an nonarbitrary way on probability considerations alone. The situation becomes even more complex, much more complex, indeed, if we ask, how much is the coin biased? 70%? 80%?, 74.873%? The probability estimates may favour different hypotheses at different stages of the experiment, and no stage is epistemically privileged.

The reason why I am questioning the validity of reasoning based on probabilistic considerations alone is not the implausibility of premises but, rather, their dependence on hidden, arbitrary assumptions. After our having 100 tails, one by one, we may infer that the probability of having the tail in the next toss is very high, for the experience suggests that the coin is biased. But we may as well infer that the probability of having the tail in the next toss is very low, for the probability of having 101 tails one by one is very low. Both inferences presuppose that the results of individual tosses are – positively or negatively – statistically dependent on each other. We may also assume the statistical independence of individual tosses and refuse to infer the probability of having the tail in the toss number 101 from the fact of our having 100 tails one by one before. On the assumption of positive statistical dependence, the more often we have tails, the higher is the probability of having one more tail. On the assumption of negative statistical dependence, the more often we have tails, the higher is the probability of finally having a head. On the assumption of statistical independence, the probability of having one more tail does not depend on how often we have had tails before.

To conclude, the techniques of probability calculus alone do not determine the way of inferring predictions from experience. It is the assumptions concerning – the positive or negative – statistical dependence or independence between the events of a certain kind that determine the choice between adopting the inductive, the counterinductive, or the apriorist rule of inference. These assumptions are logically on a par. The question is whether they are empirically on a par. Suppose we try to test them. In order to do this, let us examine whether the frequency of having the tail immediately after having 100 tails one by one is high – which apparently would support the rule of induction – or low – which would suggest the validity of the counterinductive strategy – or average – which would please the apriorist. The suitable experiment would consist in tossing the coin long enough to have, say, 100 subsequences of successive tails, and recording the results of the subsequent toss, after every such a subsequence.

Apart from the practical considerations – how long should we toss the coin? – such a test could not decide the question for purely theoretical, conceptual reasons. This is because the length of the subsequences in the experiment are established on an arbitrary basis. Suppose in the sequence, say, of 10^{10} tosses there will be 100 subsequences of 100 successive tails after which we will have, say, 80 subsequent tails, the counterinductivist need not give up. S/he may be lucky enough to find out that after we carve the initial sequence in subsequences, say, of $100+n$ tosses, the evidence will favour his/her thesis. If s/he does not succeed, s/he may point to the statistical insignificance of the result for some sufficiently big n (in the extreme, $n = 10^{10} - 101$) and require the experiment to continue. The apriorist, in turn, need not even embark upon the dispute.

The above suggests that all attempts at founding the logic of induction on the probability calculus alone have to start from some assumptions that cannot be inductively justified. I develop this point further in my (1994). What I want to suggest is that the logic of science is not the logic of induction. If we take our problem of having 100 tails one after the other as a simplified model of a scientific problem, no calculation of the probability of having the tail in the next toss can be a model of its scientific solution. What scientific rationality requires is formulating of an explanatory hypothesis entailing an independent test of the alleged bias of the coin. For example, determining the position of the coin's center of mass. Without such a hypothesis ("The centre of mass of the coin does not coincide with its geometrical centre") and without any independent evidence in its favour, predictions ("The next time will be the tail again") made on a (narrowly conceived) inductive basis (the probability considerations alone) can be said, following Hume, to have resulted from just a habit. They can be useful, but they represent a prejudice rather then a piece of (hypothetical) knowledge. On the other hand, independently testable explanatory hypotheses would suggest which of the alternative meta-inductive assumptions – the inductive, counterinductive, or apriorist – is plausible in the case in question. Thus it is the explanatory hypotheses which lend credibility to predictions made on an inductive basis.

Without explanation, there is no induction. It follows then that a genuine explanation has to yield some novel predictions. Consequently, *ad hoc* explanations – which in the foregoing have appeared to be the real target of van Fraassen's challenge – are not explanations at all.

4. Inference to the best explanation

The view that explanatory considerations play a crucial role in inductive, or ampliative, reasoning has perhaps most fully been advocated by Peter Lipton (1991). The idea of inference to the best explanation can be traced back to Charles Peirce; and in a realist's formulation is found in (Harman 1965). In the following, I shall sketch Lipton's view and offer a few amendments to it.

The principle of inference to the best explanation, in Lipton's formulation, says that "the explanation that would, if true, provide the deepest understanding is the explanation that is likeliest to be true" (*ibid*, p. 63). This principle is offered as a solution to the problem of induction. Lipton argues that this solution is free from the notorious disadvantages of its alternatives. Among other things, inference to the best explanation avoids paradoxes of confirmation that plague, as he calls it, the instantial model of confirmation – or, as I would put it, any model of induction that rests on the idea of (gradual) confirmation; it is more effective in comparison with the falsificationist method for it enables us to eliminate some unfalsified hypotheses; it accounts for the context of discovery. The principle is also reflexive: it constitutes the best explanation of our inductive practice, insofar as it points to its comparative merits with regard to its alternatives. In contrast, neither the instant-confirmation model of induction can confirm itself, nor can the hypothetico-deductive (the falsificationist) model falsify the principle of inference to the best explanation.

The essential feature of explanation, in Lipton's account, is its contrastiveness. Following van Fraassen (1980, pp. 126-129) and Garfinkel (1981, pp. 28-41), an explanation is an answer to a question of the form "Why *P* rather than *Q*?" rather than to a why-question *simpliciter*, even if *Q* is often only implicit. "To explain why *P* rather than *Q*, we must cite a causal difference between *P* and not-*Q*, consisting of a cause of *P* and the absence of a corresponding event in the history of not-*Q*" (Lipton 1991, p. 43).

The structure of contrastive explanation is illustrated by Lipton's reformulation (*ibid*, pp. 79-88) of the classical Hempel's story (Hempel 1966, pp. 3-8) of Semmelweis's discovery of the cause of childbed fever. In the hospital where Semmelweis worked, there were two maternity divisions which considerably differed in the rate of contraction of a mysterious and terminal disease called childbed fever. Semmelweis put forth and subsequently rejected several hypotheses on the basis that a putative cause of the disease either did not mark any difference between the divisions or that its elimination did not affect the difference in mortality. Lipton argues, contrary to Hempel, that the evidence to the disfavour of these hypotheses does not falsify them, unless we adopt an otherwise implausible *ceteris paribus* approach (*ibid*, pp. 91-93). Similarly, Semmelweis's reasons for accepting the final hypothesis (of cadaveric matter) are claimed to be hardly describable in terms of instantial confirmation. Instead, the evidence is analyzed as a suitable kind of contrast.

One of the amendments of Lipton's account I want to suggest is an explication of the "best" in the phrase "the best explanation". It is not that Lipton himself failed to give any explication. His observations concerning explanation-comparison are impressively extensive and deep. First of all he makes a distinction between the "likeliest" and the "loveliest" explanation. It is the latter which he regards the best. Next, he illustrates different aspects of the concept with numerous examples. The scope and intuitive plausibility notwithstanding, his analysis suffers from the lack of a comprehensive summary – other than just listing a few explanatory factors: unification, mechanism, and precision (*ibid,* p. 124). In my attempt at formulating some explicit criteria of loveliness, I shall make use of conceptual devices developed by van Fraassen.

It may seem ironic that I borrow some ideas from van Fraassen's account of the pragmatics of explanation to support the denial of what he denies: that explanatory power is an epistemic virtue. In doing this, I have no mischievous intentions. I simply acknowledge the powerfulness of the conceptual machinery he has invented and try to apply it for my own purposes. For van Fraassen, an explanation is a good answer to a why-question. In his analysis of why-questions (van Fraassen 1980, pp. 141*ff*), van Fraassen identifies a why-question Q with a triple $\langle P_k, X, R\rangle$ of, correspondingly, the *topic, contrast-class* and *relevance relation* of the question. The topic P_k of Q is the proposition which is obtained from Q by deleting the word "why" and the question mark. The contrast class $X = \{P_1,..., P_k,...\}$ consists of alternatives to the topic and the topic itself. The relevance relation R picks up propositions that are relevant to the couple $\langle P_k, X\rangle$ in the context. Both X and R depend on the context. The context include a certain body K of background knowledge (theory and factual information).

Now, B is a *direct answer* to Q if and only if there is some proposition A such that A bears relation R to $\langle P_k, X\rangle$ and B is true if and only if P_k; and for all $i \neq k$, *not* P_i; and A is true. A is called the *core* of answer B, and B is abbreviated to "Because A". In offering an explication of the concept of loveliness of explanation I shall not pursue van Fraassen's line of evaluating possible answers to a why-question. My reason is that van Fraassen is interested in giving an account of how science is applied to produce good explanations. Consequently, his criteria of goodness of explanation assume a fixed state of knowledge (which is a part of a context). In contrast, my present task is giving an account of how to evaluate (rival) hypotheses on the basis of how good the explanations they provide are. Consequently, the evaluative criteria I am going to offer have to permit comparisons across alternative hypothetical states of knowledge.

The intuitive idea is that the greater the explanatory power of a hypothesis the greater number of possible applications in the industry of explanation it has. I propose the following definition: a hypothesis H is applied in giving an explanation relative to some background knowledge K if there is a couple $\langle P_k, X\rangle$ such that P_k and for all $i \neq k$, *not* P_i and H together with K: (i) determines a relevance relation R for $\langle P_k, X\rangle$ and (ii) favours a direct answer to $Q = \langle P_k, X, R\rangle$. To use Hempel-Lipton's example, Semmelweiss's hypothesis of cadaveric matter established the relevance of patients having been examined by medical students who performed their examinations directly after performing autopsies and favoured the answer "Because the cadaveric matter is frequently introduced to patients' blood-stream in the course of examinations performed by students" to the question "Why the mortality rate is high in the First Division and low in the Second Division?". As we know, today we are able to give a better explanation, in terms of infection by micro-organisms. Why is this better?

According to the above stated one-many-relation (of a part of *explanans* to *explanandum*) requirement, if a hypothesis H is to be explanatory at all, it is necessary that it is possible for H to do the same job for some other $\langle P_k', X' \rangle$. Let us call the class of all couples $\langle P_k, X \rangle$ for which a hypothesis H, together with some background knowledge K which does not entail that P_k is not true, determines a relevance relation R and favours a direct answer to $Q = \langle P_k, X, R \rangle$, the class of possible potential explanatory applications of H relative to K, CPEA(H, K), for short. (An explanatory application of H is a potential one *simpliciter* if, in addition, P_k is, in the light of K, known to be the only true proposition in X. An explanatory application of H is actual if, in addition, both H and the direct answer to Q are true.). The intuition behind the one-many-relation requirement is that if CPEA(H, K) is a unit class, the hypothesis H is *ad hoc* and hence is not genuinely explanatory. This very intuition leads us to think that the wider CPEA(H, K) is, the greater the explanatory power of the hypothesis in question is. Thus we obtain a criterion for the comparative assessment of the explanatory power of alternative hypotheses: if the CPEA(H, K) is included in CPEA(G, K), then the hypothesis G is of greater explanatory power than the hypothesis H. Now, whatever the classes of possible potential explanatory applications of Semmelweiss's hypothesis of cadaveric matter and our theory of infectious diseases are, the latter is more inclusive than the former.

Now, let us reflect for a moment on the relation of the criterion on offer to the explanatory factors which have been listed by Lipton: unification, mechanism, and precision. First, if a hypothesis H represents a unification of some hypotheses G_1, G_2, then CPEA(H, K) obviously includes CPEA(G_1, K) and CPEA(G_2, K). Second, if a hypothesis H reveals some mechanism, it will provide some hints towards the invention of new possible potential explanatory applications of the hypothesis in question and thereby widening CPEA(H, K). Third, if a hypothesis H is more precise than some other hypothesis G, and all possible potential explanatory applications of G are also possible potential explanatory applications of H, then H allows us to split at least some elements P_1, P_2... of at least some classes of contrast X_1, X_2... in CPEA(G, K) and thereby form some possible potential explanatory applications for H which were absent from CPEA(G, K). Thus we can see that our formal criterion captures the intuitions behind Lipton's analysis. Consequently, I claim that the best, or the "loveliest" explanation is the explanation which is entailed by the hypothesis of the most inclusive class of possible potential explanatory applications in comparison to the classes of possible potential explanatory applications of its rivals.

5. Explanation, testing, and cross-background-knowledge comparisons

As mentioned above, Lipton provides some arguments to the effect that his best-explanation account fares better than its major rivals: the instantial model of confirmation and the falsificationist model. In what follows I shall point to some further advantages of this kind of approach. To start with, however, I shall put into question his very statement of the "inference to the best explanation" principle. He says, I repeat, that "the explanation that would, if true, provide the deepest understanding is the explanation that is likeliest to be true" (*ibid*, p. 63). In Sec. 3 I criticized the attempts to model inductive reasoning on probability considerations alone. Apparently, Lipton's formula fills the gap in the Bayesian approach by supplementing it with explanatory considerations, which I have claimed to be necessary. Still, I claim, the range of applications of the concept of likelihood is limited to specific types of situations.

What I have in mind is that to define a probability distribution we have to have, first, a fixed probability space, i.e. the space of all possible hypotheses in the given context of inquiry *and*, second, a good explanation as to why the distribution we offer is not arbitrary. For example, in an ethological theory, in the face of a danger, an animal may either try to escape or attack. Thus, in an attempt at answering the question "What will such-and-such animal do when threatened in such-and-such way?" we may define the probability space of two hypotheses: "It will try to escape" and "It will attack". Next, relying on further details of the theory in question, we may choose an appropriate probability distribution over this space. The probability of either hypothesis is estimated then on the assumption that the explanations provided by the ethological theory in question are true. We may say that these explanations, if true, provide a plausible – perhaps the most plausible – probability distribution over the space of the hypotheses under consideration. The quality of these explanations, however, has nothing to do with probability estimates of these explanations themselves. In this view, Lipton's use of "likeliest" in this context is either wrong, if intended to suggest comparisons in terms of probability, or figurative, if not. Taking into account Lipton's distinction between the "likeliest" and "loveliest" explanation, I intend to take the formula under discussion as a slip of tongue – otherwise, the "best", i.e. the "loveliest" explanation would inevitably collapse into the "likeliest" one, which would clearly run counter to Lipton's intentions.

I shall defer from suggesting a plausible replacement for "likeliest" in Lipton's formula until I make some further points. First, note that any nonarbitrary probability distribution over any space of hypotheses has to be defined relative to some background knowledge. Hence any revision of background knowledge may affect the probabilities, while the revision itself is not accountable in terms of the probability comparisons of the rival systems of background knowledge – unless we introduce the hardly coherent notion of super background knowledge. It follows, then, that if the probability estimates were the only means of comparing theories, any comparison would be valid only if made against the same background knowledge, and no means of comparison between rival systems of background knowledge would be available. This I would call the strong "incommensurability fallacy" of Bayesianism.

There is also a weak form of incommensurability fallacy I am inclined to ascribe to Popperianism. Popper in his (1963) offers three "conditions of the growth of knowledge", i.e. three criteria for making evaluative comparisons of rival hypotheses. These are: simplicity, empirical content, and relative empirical success. The trouble is that these criteria only work when applied against a shared background knowledge. As far as I know, Popper has not explicitly offered any criteria for evaluative comparisons of rival systems of background knowledge. In my (1994) I claimed that Popper did offer some implicit ones and I tried to spell them out. (Therefore I call this a weak form of incommensurability fallacy). Still, I think that I am now in a position to give a more powerful account of cross-background knowledge comparisons in terms of Inference to the Best Explanation.

To start with, note that the comparisons of CPEA(H_i, K_j) are possible not only for different i's and a fixed j but also for a fixed i and different j's. In other words, we can make sense of the concept of explanatory power of a background knowledge relative to a fixed hypothesis. The concept can be generalized to that of explanatory power of a background knowledge relative to a stock of hypotheses in all cases in which the result of background-knowledge comparison does not depend on the choice of a hypothesis from the stock in question. This idea is somewhat related to the one I offered in my (1994) to amend Popper's account, but it is now put in much more exact terms. It is also possible to vary both *i*'s and *j*'s. This enables us also to relate the idea on offer to the Lakatosian question of comparison of alternative scientific research programmes (with "hard core" as a background knowledge) or Laudan's (1977) proposal of solving Duhem's problem in terms of problem-solving efficiency. One merit of the present proposal is the avoidance of Lakatos's obscure concepts of "progressiveness", especially "heuristic progressiveness", of a research programme as well as of Laudan's intractable problems of problem individuation and problem weighing. Later, I shall offer some sophistications to this relatively simple idea of comparisons with respect to explanatory power. Before this, let me explore its immediate consequences.

The criterion on offer imposes a partial ordering on the set of pairs $\langle H, K\rangle$ of competing hypotheses and systems of background knowledge in a given domain. This feature lends support to the advice of *theoretical pluralism* which has been advocated most notably by Feyerabend and Lakatos, while avoiding accompanying Feyerabend's epistemological anarchism or Lakatos's ambiguities of conception. The point is that explanatory power (of a hypothesis relative to a background knowledge) is but a promise of providing us with a deeper understanding, *if* the hypothesis in question is true. This "if" points to the need of testing. On the best-explanation account, a test of an explanatory hypothesis is neither an attempt at its confirmation nor an attempt at its falsification. Rather, a test is an attempt at broadening or narrowing down the class of possible potential explanatory applications of the hypothesis in question and subsequently comparing it to the corresponding classes of its alternatives.

Basically, there are two ways in which a possible potential explanatory application $\langle P_k, X\rangle$ can fail to be an actual explanatory application of a hypothesis H relative to a background knowledge K. First, the topic P_k need not be true, or needs not be the only true proposition in the contrast class X. Second, H may favour an untrue direct answer to $Q = \langle P_k, X, R\rangle$, or H may itself be untrue. For the purpose of simplicity, assume that both the topic P_k of the why-question under consideration and the H-favoured direct answer to it are singular propositions. Now, an attempt at finding out whether the topic P_k of a possible potential explanatory application of H is true can, in principle, be described as a test of whether a prediction to the effect that, in certain circumstances, P_k, and not any other P_i from X, will hold. The result of such a test will be an "instant-confirmation" or "disconfirmation" of *H*. Note that disconfirmation of this sort cannot be described as "falsification" unless we invoke the concept of a *ceteris paribus* clause. Concerning the assessment of an *H*-favoured direct answer to *Q*, its test will involve, according to Lipton's account, an attempt to establish whether the presence of the factor mentioned in this answer represents a suitable contrast between the topic P_k and other members of the contrast class *X*.

Let me give an example to illustrate how the criterion on offer works. Consider the question "Why does the anomaly (with regard to Newtonian celestial mechanics) of Mercury's perihelion occur?" Its contrast is "...rather than not". Few tentative answers – e.g. the one based on the hypothesis of the existence of a hitherto unknown planet Volcano – were rejected on the available evidence. The first plausible answer was drawn from general relativity theory which has been offered in Newtonian celestial mechanics stead. Later it was discovered that a Newtonian answer was still possible: the appropriate explanatory hypothesis assumed a suitable kind of irregularities of solar mass distribution. In the Popperian-Lakatosian account, the hypothesis is to be rejected on the grounds that it is a *post hoc* hypothesis. What is wrong with *post hoc* hypotheses? Let *H* be the hypothesis under discussion, *K* be the system of background knowledge which includes, among other things, Newtonian mechanics, *K'* be the result of the replacement of Newtonian mechanics in *K* by the general relativity theory. Now, the question about Mercury's perihelion is an explanatory application of *H* relative to *K* as well as of *K'* relative to itself. But the explanatory power of the latter is greater, for it has, in addition, some other explanatory applications, e.g. the question "Why does the red shift phenomenon occur?"

So far, no special novelty is introduced in comparison to Lipton's account of testing, apart from, possibly, some conceptual clarification. Our conceptual machinery, which we have borrowed from van Fraassen, permits us, however, to introduce some refinements as mentioned earlier.

6. Explanation and presupposition

A why-question presupposes that (i) in its contrast-class the topic, and only the topic is true, and (ii) at least one proposition that bears relevance relation to its topic and contrast-class is true (cf. van Fraassen 1980, p. 145). Apparently, the truth of (i) is just a logical question of the choice of a suitable contrast-class. This is not so, however. In Semmelweiss's day a suitable contrast to, say, "A patient will die by such-and-such time" would certainly be "A patient will live at least to such-and-such time". In our days the contrast is not so obvious, considering the possibility of artificially maintaining the vegetative functions of the organism of a patient who is otherwise unable to live. It appears then, that (i) can be false even if no logical mistake in defining the contrast-class is made. For it is always possible that we will encounter a novel empirical fact that cannot be accomodated with the logic of our language without a revision of that language. That (ii) can be false is more obvious. I propose to include presuppositions of a why-question to the background knowledge against which the problem of giving an explanation is considered. A revision of background knowledge may involve the rejection of presuppositions of a why-question and thereby the rejection of the why-question itself.

Turning back to our previous definition of explanatory power, we may counter-intuitively conclude that whenever we discover or conjecture that some presupposition of a why-question is false we are faced with some explanatory loss, "Kuhn's loss", indeed. To avoid this, I propose the following amendment to our previous definition: in cross-background knowledge comparisons, one has to exclude from CPEA(H, K) all the possible potential explanatory applications of H relative to K (some of) whose presuppositions are rejected on the grounds of the competing system of background knowledge.

As it has been said, presuppositions of a why-question, and hence the why-question itself, can be rejected on the assumption of some background knowledge. The question arises as to whether the reverse may happen: whether the need of a revision of background knowledge can arise from the assumption of the falsity of some presuppositions. A possible positive answer would establish some further symmetry – in addition to that of the comparisons with regard to explanatory power – between the epistemic status of hypotheses and background knowledge: a request for explanation could motivate the search for a new hypothesis as well as for a revision of background knowledge.

The answer I want to suggest is, indeed, positive albeit qualified. A prolonged failure in finding a plausible answer (among propositions that bear relevance relation to its topic and contrast-class) to a why-question – that is, a serious doubt in the presupposition of type (ii) – *can* be taken as a piece of evidence in disfavour of the current background knowledge. On the other hand, one may always tend to keep background knowledge intact and blame the scientists for the lack of ingenuity instead: there is always a possibility of inventing a hypothesis which will affect the relevance relation and hence the class of possible answers. This is Kuhn's well-known dilemma: in the face of recalcitrant puzzles one can either insist on maintaining a paradigm come what may or one can promptly embark on extraordinary research. In the present account the dilemma is as forceful as Kuhn has had it. The important difference between Kuhn and me is that rival paradigms – once a rival comes about – are, in my proposal, "commensurable" with regard to explanatory power. On the other hand, this "common measure" imposes a partial rather than a linear ordering on the set of competitors. Consequently, there is still room for – if only temporarily – unresolved disagreement.

The questioning of the presupposition of type (i) reveals a more complex pattern. To illustrate this, suppose we have two different why-questions, $Q = \langle P_k, X, R\rangle$ and $Q' = \langle P_k', X', R'\rangle$ and two rival hypotheses, H and H' which favour a direct answer "Because A" to the question Q and "Because A'" to the question Q', respectively. Suppose next that, given some background knowledge K, P_k and P_k' are confirmed and both H and H' are hypotheses of maximal explanatory power in the domain, relative the background knowledge K, i.e. they are maximal elements in the partially ordered, with regard of the relation of explanatory power relative to the K, set of hypotheses on offer. Suppose finally that A entails $\neg P_k'$ and A' entails $\neg P_k$. In other words, accepting any of the rival hypotheses would commit us to rejecting a presupposition of a why-question answered by the other hypothesis, the presupposition which otherwise is accepted on the evidence at hand, in the light of K. In such a situation we must either reject both the rival hypotheses, or the evidence, or else revise the background knowledge. Rejecting the evidence amounts to the charge of fraud or serious error on the part of the experimenter, which is always possible. But taking evidence seriously, we are left with two possibilities. Sometimes, the first horn of the dilemma (or trilemma) is highly implausible, especially when $H' = \neg H$. This may strongly suggest the need for a revision of background knowledge. In particular, the to-be-revised background knowledge may contain – sometimes only implicitly – some untrue presuppositions of H, in the technical sense of the term, in which a presupposition of a proposition H is a necessary condition for it to have a truth-value (cf. van Fraassen 1968). In such a case, a suitable revision of the background knowledge allows us to reject both H and $\neg H$ without violating logic. Still, the search for alternatives *within* K need not be unreasonable. Which way is better, we can judge – using our criterion of explanatory power – only after a few alternative solutions are put forth. Again, the criterion imposes a partial rather than a linear ordering on possible solutions and hence there is still room for – possibly only temporary – disagreement.

A good example of this kind of scientific problem is the request for an explanation of the null result of the Michelson-Morley experiment. The tentative explanation in terms of Earth's drag of ether was in conflict with tentative explanations of some other phenomena in terms of etheric wind. After decades of unsuccessful attempts at resolving the conflict with different variants of the partial drag hypothesis, special relativity theory explained the Michelson-Morley effect and denied, at the same time, a shared presupposition of rival ether hypotheses: that the movement of a body relative to ether can be empirically detected. Prior to special relativity, Lorentz's contraction hypothesis, which did not involve a revision of background knowledge, was tried. But after rejecting some of its intended explanatory applications, ones which concerned the unsuccessful prediction of some electromagnetic phenomena, it appeared that the special relativity theory's explanatory power is greater than that of Lorentz's hypothesis.

Concluding, the best-explanation approach accounts for various aspects of comparative assessment of scientific theories or even the whole bodies of theories, e.g. alternative systems of background knowledge. Lipton's principle of contrast, in comparison to alternative instant-confirmation or falsificationist principles, makes hypotheses more vulnerable to test. On the other hand, contrastive evidence is conceived as a much "softer" kind of evidence than "confirmation" or "falsification". Thus the best-explanation approach lays even more stress than its alternatives on the conjectural character of scientific knowledge. It provides an account of cross-background knowledge comparisons in more exact terms than its rivals. It accommodates the principle of theoretical pluralism without any element of epistemological anarchism. In particular, it admits the possibility of prolonged rational disagreement without commitment to the incommensurability thesis. It makes some hints towards accounting for the mechanism of vast conceptual changes in terms of rejecting some – sometimes only tacit -presuppositions of a range of alternative hypotheses. In the light of the above discussion, explanatory power, contrary to van Fraassen's view, is not a purely pragmatic virtue. We compare our hypotheses and whole bodies of theories first and foremost with respect to their explanatory power. Thus, if science is a cognitive enterprise, explanatory power is a cognitive virtue. On the other hand, the present account inherits those features of van Fraassen's prototype that permit the capture of the pragmatics of explanation: possible relativization of explanation to some background knowledge, in particular to one's personal knowledge. This, however, is not the primary aim of the present analysis.

7. Inference to the Best Explanation and Common Cause Principle

Inference to the Best Explanation Principle is not only a means of assessment of scientific theories but also of methodological rules. To illustrate this claim, I shall briefly discuss the Common Cause Principle.

The Common Cause Principle states, according to Reichenbach (1959), that every statistical correlation is required to be given an explanation in terms of a common cause of the correlated phenomena. The Common Cause Principle when applied to EPR-type quantum correlations amounts to a postulation of hidden variables, which is regarded as implausible by most quantum theorists. Van Fraassen (1980) takes this as evidence against the latter principle in question. Further, since the Common Cause Principle is, in his view, but a version of the Inference to the Best Explanation Principle, this is also put into question. He admits the heuristic value of both the principles but rejects them as universal methodological rules.

Of course it is the scientist and not the philosopher who decides the questions of quantum theory. Consequently, even though the question of hidden variables might be viewed as an unsettled one, the philosopher must be prepared to reject the universal applicability of the Common Cause Principle. This should not to be done, however, solely on the basis of a single recalcitrant piece of evidence which is thought to be provided by the case of the failure of hidden variable theories. Instead, an attempt should be made to arrive at a more general theory of the conditions of the applicability of the principle in question.

One proposal concerning a qualified approach to the Common Cause Principle was given in (Brown 1982). His account is illustrated by a story of an exploding church. The story tells that one evening all the members of a church's choir, for accidental and independent reasons, were late for a rehearsal. And this was a lucky coincidence, because the church exploded at the exact time that the rehearsal was due to begin. The journalist who told the story concluded by asking the rhetorical question as to whether the members of the choir had been saved by God's interference. God's interference hypothesis was intended here as an instance of the application of the Common Cause Principle. Brown suggests, reasonably enough, that this is not the right sort of application of the principle. Unfortunately, his positive offer with regard to this question is entirely exhausted by the formula that in doubtful cases of application of the Common Cause Principle we should rely on the usual standards of rationality.

This sort of answer is somewhat puzzling. Does not the Common Cause Principle itself belong to the standards of rationality? And what does it mean to apply the standards of rationality to the standards of rationality? Is this to imply that there is a hierarchy in the standards of rationality? Brown has left these questions unanswered. I shall try to answer the last question in the positive and indicate the superiority of the Inference to the Best Explanation Principle.

Consider the exploding church story from the perspective of this principle. There is a natural explanation of the lifesaving of the members of the church's choir: *X* was late for the reason *p*, *Y* for the reason *q*, etc. The journalist suggested the God's intervention hypothesis as an alternative explanation which, philosophically speaking, satisfies the Common Cause Principle. According to the contrastive account of explanation, the question arises what is the contrast-class for the topic "all the members of the church's choir were saved from the explosion by causing them to have been late for the rehearsal". Suppose for the sake of argument, that we have chosen as the contrast-class the proposition of the kind "these-and-these the members of the church's choir were saved, while those-and-those were killed by the explosion". Now, does the hypothesis of God's intervention provide an explanation why all the members of the church's choir were saved rather than all but a hypocrite *Z* who entered the choir to save appearances and was, in fact, a godless and wicked person? Let us try another contrast. Does the hypothesis under discussion explain why they were saved by causing them to have been late rather then by preventing the bomb from exploding? The God's intervention hypothesis is of a very low, if not null, explanatory power, for He is, by His very nature, unpredictable. He may save the lives of His favourites (who knows who they are?) or, to the contrary, call them to Him. He may do this or that in any conceivable or inconceivable way, almighty as He is.

Note, that the question is not really the one of inductive reasoning, as traditionally conceived. The explanatory weakness of the hypothesis of God's interference has nothing to do with having too few churches exploded and the members of choirs saved. Nor with the moral unacceptability of performing relevant experiments. The hypothesis under discussion does not explain anything because we cannot, in the context of our knowledge, point to a suitable contrast-class for our topic.

For this reason a set of usual causal explanations, separate for every single member of the choir, of his or her having been late, is to be preferred. Thus, the Inference to the Best Explanation Principle sets some limitations on the scope of the applicability of the Common Cause Principle. Consequently, the failure of attempts at applying this principle in the search of explanation of some quantum correlations does not undermine the principle itself, let alone the Inference to the Best Explanation Principle. By this very principle, no causal explanation of EPR-type phenomena is available, at least at present. In other cognitive contexts, the best-explanation approach can even help to fix the best general pattern of an explanation in the given context: causal, intentional, functional, etc. This is one more revision, or expansion, of Lipton's account.

8. Explaining the success of science

There is a meta-level application of the Inference to the Best Explanation Principle, which is sometimes called "the Ultimate Argument" in favour of realism in the philosophy of science. The argument says, in Putnam-Boyd (cf. Putnam 1978, p. 73) formulation, that realism is the only view on which the success of science need not be conceived as a miracle. In other words, realism provides the only, or the best explanation of the success of science.

This argument is questioned not only by the antirealist. Some realists also tend to doubt its plausibility. In my view, their attitudes are unjustified and mistaken. Lipton (1991, p. 172), for example, takes Putnam's maxim as redundant, for the Inference to the Best Explanation Principle does already imply that the theories accepted on its grounds are likely to be true. There is possibly, however, another reading of the principle. Taken in its generality, the principle is a guide for acceptance or, better, preference of the theories or hypotheses on offer. The antirealist is free to divorce acceptance from the likelihood of being true and, like van Fraassen, marry it to the likelihood of being empirically adequate, i.e. true as far as the observable entities are concerned. Consequently, van Fraassen is free to adopt the Inference to the Best Explanation Principle in an antirealist version – and indeed he does this – in which the principle is a guide to empirical adequacy rather than truth, i.e. truth that goes beyond the domain of the observable. Thus the ultimate argument concerns the question of what is the appropriate reading of the Inference to the Best Explanation Principle. The argument can then be rephrased in the following way: on the

Inference to the Best Explanation Principle, applied to the explanation request concerning the success of science, the appropriate reading of this very principle, in its applications to scientific explanation, is that this principle is a guide to truth.

In this reformulation of the ultimate argument I have purposely avoided two doubtful notions. First, as I have already said, I question Lipton's suggestion to the effect that the best explanation is the one *likeliest* to be true. Second, I find Putnam's formulation of the realist position in terms of belief that our theories are *approximately* true implausible. My reasons for avoiding probabilistic notions have been already mentioned. My reasons for avoiding the concept of approximation to truth is that I doubt whether this concept really makes sense. I do not believe that there is such a thing as the Truth, a fixed goal we get ever closer to in the course of inquiry. Rather, I take truth to be a regulative idea, which points to a certain direction, towards an infinitely distant aim of inquiry (cf. Watkins 1984, p. 124). Consequently, I take the Inference to the Best Explanation Principle to be a *guide* to truth, in the sense other than that of a measure of the likelihood or distance from the truth.

I am taking here the success of science for granted, although I must admit that this has been challenged, most notably by Feyerabend, and the very concept of success is not at all clear. I shall confine the present discussion to the comparison of two alternative explanations of the assumed success of science with respect to their explanatory power. The most articulate alternative to the ultimate explanation is van Fraassen's (1980) pseudo-Darwinian account. In his view, the explanation is that only successful theories survive while unsuccessful ones are extinguished. Next, a theory is successful if it can plausibly be claimed to be empirically adequate, which is something less than true or approximately true. It is claimed then that empirical adequacy is the only, or the basic, adaptative virtue in the process of the selection of theories. The very concept of empirical adequacy rests on the questionable distinction between the observable and the unobservable. Although the distinction is drawn in a way that avoids the notorious shortcomings of its logical positivists' namesake, it still has many puzzling consequences. I joined the choir of its critics in my (1991, 1995). For the present purposes, however, we can take this distinction for granted and ignore its peculiarities.

One counter-example of van Fraassen's pseudo-Darwinian account is Chinese medicine. The latter seems to be therapeutically so effective that many physicians are persuaded to practise some of its methods. At the same time, these very methods are viewed with suspicion because of their non-scientific character. It is not that there is no theory which can account for the success of, say, acupuncture. In fact, there is one, the theory of life energy transmission channels. This theory, given the success of acupuncture, is bound to be considered, in van Fraassen's eyes, as an empirically adequate one. Yet the theory in question is not accepted as a scientific one. This is evidence that empirical adequacy of a theory is not sufficient for its survival *in science*, even in the absence of a serious competitor. It is the *practice* based on an empirically adequate theory that survives. Theories, in contrast, seem to be subjected to some additional selective factors.

One may be tempted to suggest that the theory of Chinese medicine cannot survive in the environment of Western science just because it comes from an alien tradition. But it is much more plausible to suppose that this theory is not accepted because of its incompatibility with other theories of Western science and this incompatibility proves that Western theories and the Chinese theory cannot all be true. This supposition accounts for a strong selective pressure whose operation can also be seen within Western tradition, in situations where we face conceptually incompatible, observationally non-conflicting, and apparently empirically adequate theories. Consider, for example, the tension between classical gravitation theory and special relativity.

It was clear almost at the outset of special relativity theory that gravitational phenomena could not be accounted for within its conceptual framework. At the same time, both the classical gravitation theory and special relativity theory had good evidential support. There was no reason to question the empirical adequacy of either of them. Yet the conceptual disparity between the two theories led Einstein to undertake the general relativity research programme and eventually to find a unified account of the phenomena in the domains of both predecessors.

This kind of striving for theoretical unification can be explained by the presence of an additional selective factor: the operation of the regulative idea of truth that guides the research. With empirical adequacy considerations alone, we would have no reason to question neither the classical gravitation theory or the special relativity theory. Thus the ultimate explanation of the success of science is more powerful, with respect to cases of this kind than van Fraassen's "penultimate" one.

This line of reasoning has been questioned in, e.g. (Newton-Smith 1989). Despite his realist position, he has suggested that the antirealist is free to claim that it is a bare empirical fact that successful unifications lead to some gains in empirical adequacy. Consequently, Newton-Smith is prepared to abandon, as he calls it, explanatory realism in favour of "modest" realism. This concession is too hasty, I think, taking into account the role of explanatory considerations in the testing of theories. Invoking bare coincidence between unifications and gains in empirical adequacy is of zero explanatory power. In contrast, invoking the regulative idea of truth explains both the strive towards unifications *and* the possible resultant gains in empirical adequacy. It is the more so that similar gains in empirical adequacy can be achieved without unifications, as is suggested by the already mentioned example of explaining Merçury's precession by the ununiform solar mass distribution hypothesis. In van Fraassen's account, there is no reason why gains in empirical adequacy due to unifications are to be preferable over the gains due to *post hoc* hypotheses. In the present account, any unification of the sort under discussion is preferable as a move from an established falsity to a possible truth.

By way of further illustration of my claim that the ultimate explanation of the success of science is superior to the pseudo-Darwinian one, let us compare the argument on offer with the exploding church case. Van Fraassen's explanation, in terms of empirical adequacy, points to the virtues of particular theories which are assessed independently from the success of other theories. Their joint success is then explained analogously to the explanation of the exploding church story in terms of giving independent reasons for being late for every particular member of the choir. In the antirealist view, possible coincident truthlikeness of supposedly empirically adequate theories is as equally inexplorable as the alleged miraculous God's intervention. This analogy is, however, misleading. The hypothesis of God's intervention hardly contributes to the understanding of natural reasons for the members' of the choir being late. To the contrary: on the assumption of God's intervention their reasons might have been different, while the effect must have been the same. The realists explanation of the success of science, on the other hand, does contribute to the understanding of why particular virtues of particular theories have proved their adaptive superiority. While the members of the choir could have been late for independent reasons, scientific theories can hardly be considered successful, each in its own domain, independently from the appraisal of other theories. For scientific theories enter into numerous intertheory relations, including presupposition relations.

To conclude, the Inference to the Best Explanation Principle can serve not only as a means of comparative evaluation of scientific theories. It can also serve as a means of evaluation of other methodological rules and even of epistemological theories. Unifying three levels of discussion – one of scientific theories, one of scientific method, and one of the aim science – the best-explanation account explains why theory comparisons with regard to theories' explanatory power are a guide to the truth.

T. Kotarbiński Pedagogic University
Al. Wojska Polskiego 69
65-762 Zielona Góra, Poland
and
Jagiellonian University. Kraków

REFERENCES

Brown, J.R. (1982). The miracle of science. *Philosophical Quarterly* **32**, 232-244.
Cartwright, N. (1983). *How the Laws of Physics Lie*. Oxford: Clarendon Press.
Feyerabend, P.K. (1979). Dialogue on method. In: G. Radnitzky, and G. Andersson (Eds.), *The Structure and Development of Science*. Dordrecht: Reidel, pp. 63-132.
Garfinkel, A. (1981). *Forms of Explanation*. New Haven: Yale University Press.
Grobler, A. (1991). Van Fraassen's metaphysical move. *International Studies in the Philosophy of Science* **5** (**1**), 21-34.
Grobler, A. (1994). Justification of the Empirical Basis. Bayesian vs. Popperian Conception of Rationality. In: G. Meggle and U. Wessels (Eds.), *Analyomen 1. Proceedings of the 1st Conference "Perspectives in Analytical Philosophy"*. Berlin: Gruyter, pp. 99-309.
Grobler, A. (1995). The Representational and the Non-representational in Models of Scientific Theories. In: W. Herfel, W. Krajewski, Z. Niiniluoto, R. Wójcicki (Eds.), *Theories and Models in Scientific Processes* (*Poznań Studies in the Philosophy of the Sciences and the Humanities* **44**). Amsterdam–Atlanta: Rodopi, pp. 37-48.
Harman, G. (1965). Inference to the Best Explanation. *The Philosophical Review* **74**, 88-95.
Hempel, C.G. (1965). *Aspects of Scientific Explanation*. New York: Free Press.
Hempel, C.G. (1966). *The Philosophy of Natural Science*. Englewood Cliffs: Prentice-Hall.
Laudan, L. (1977). *Progress and Its Problems*. Berkeley: University of California Press.
Lipton, P. (1991). *Inference to the Best Explanation*. London: Routledge.
Newton-Smith, W. (1989). Modest realism. In: *PSA 1988* **2**, 179-189.
Popper, K. (1963). *Conjectures and Refutations*. London: Routledge.
Popper, K. (1972). *Objective Knowledge*. Oxford: Oxford University Press.
Putnam, H. (1978). *Meaning and the Moral Sciences*. London: Routledge.
Reichenbach, H. (1959). *Modern Philosophy of Science*. London: Routledge.
Salmon, W. (1971). *Statistical Explanation and Statistical Relevance*. Pittsburgh: University of Pittsburgh Press.
Salmon, W. (1984). *Scientific Explanation and the Causal Structure of the World*. Princeton: Princeton University Press.

Van Fraassen, B. (1968). Presupposition, Implication, and Self-reference. *Journal of Philosophy* **65**, 136-152.

Van Fraassen, B. (1980). *Scientific Image*. Oxford: Clarendon.

Watkins, J. (1984). *Science and Scepticism*. Princeton: Princeton University Press.

Poznań Studies in the Philosophy of the Sciences and the Humanities
2000, Vol. 72, pp. 89-103

Piotr Giza

"INTELLIGENT" COMPUTER SYSTEMS AND THEORY COMPARISON

ABSTRACT. The article attemps to answer the question of whether, and to what extent, contemporary computer systems designed to make scientific discoveries are capable of comparing and deciding between rival theories and hypotheses that exist in a certain domain of science. I analyze the way these systems perform their task and show the differences between computer reasoning and human thought. I conclude with the claim that computer systems, even the most "intelligent" ones, require some additional *ad hoc* heuristics that introduce human intuition, but these systems more efficiently analyze the vast amounts of information necessary to compare theories.

1. Introduction

In the 1970s a few AI researchers in the USA turned their attention to a domain often considered the realm of genius – scientific discovery. The vast majority of this work, however, has focused on empirical discovery, and much of the effort has been concerned with the induction of numeric laws from the history of physics and chemistry such as the ideal gas law, Coulomb's law, Snell's law of refraction, Black's law of specific heat, Gay-Lussac's law of combining volumes or Canizzarro's determination of relative atomic weights. The most famous computer systems performing these sorts of tasks include Gerwin's function induction system[1], Langley, Bradshaw, and Simon's BACON[2], Żytkow's FAHRENHEIT[3], or Nordhausen and Langley's IDS[4].

The empirical discovery systems mentioned share some common heuristics modelled on real-life behaviour of scientists in a laboratory, and not only are they very useful in detecting regularities in "raw" data,

---

1 (Gerwin 1974).
2 (Bradshaw, Langley and Simon 1980).
3 (Żytkow 1987).
4 (Nordhausen and Langley 1986).

some of them are capable of gathering data by way of conducting simple experiments (Langley and Żytkow 1989). Empirical discovery often occurs in the early stages of a field's evolution, before scientists have acquired much knowledge of the domain. As a result, it seems likely that general, domain-independent heuristics plays a more central role in this task than in the process of theory formation and revision. It is for this reason that discovery systems are philosophically interesting both as descriptive and normative accounts of scientific discovery. Nonetheless, these systems do not deal with what are normally considered scientific theories or theoretical entities and, in consequence, are not concerned with theory choice.

Recently, however, some new discovery systems have been created whose task is to reveal the "hidden" structure of matter in chemistry or physics. They include Langley, Simon, Bradshaw and Żytkow's STAHL and DALTON[5], Rose's REVOLVER[6], or Fisher and Żytkow's GELL-MANN[7]. All of them, by dealing with structure not available by direct observation or measurement, are bound to propose, assess and compare hypotheses about that structure that deserves the name "theories" much more than hypotheses involved in the first class of systems.

As a concrete example of a computer system used for theory construction, comparison and choice I shall examine GELL-MANN designed in 1990 at Wichita State University by a group of researchers from Automated Discovery Laboratory[8]. The program's task was to analyze elementary particle data available in 1964 and formulate an hypothesis (or hypotheses) about a "hidden", more simple structure of matter, or to put it in contemporary terms – the discovery of quarks. My choice is partly motivated by the fact that GELL-MANN demonstrates several new capabilities not available in the earlier systems that discovered hidden structure. Moreover, it deals with research at the frontier of science which makes its methodological analysis more interesting.

Section 2 contains a very brief exposition of the state of elementary particle physics in the early 1960s and the way the quark hypothesis was arrived at as a result of symmetry considerations.

In Section 3 I show how GELL-MANN proceeds to discover quarks.

---

[5] (Langley, Simon, Bradshaw and Żytkow 1987).

[6] (Rose 1989).

[7] (Fisher and Żytkow 1990).

[8] The system takes its name from Murrey Gell-Mann the first physicist who suggested, in 1963, the existence of a hidden structure in particles then widely considered "elementary".

Finally, in concluding Section 4, I make an attempt to highlight the differences in the ways theoretical physicists and GELL-MANN arrived at the quark hypothesis. I argue that they involve different levels of theory comparison and assessment.

2. How Quarks Emerged in Theoretical Physics

Until the mid-1930s the picture of the building blocks from which all matter was made had been clear and simple. Atoms consisted of a nucleus and electrons orbiting around it. The nucleus, in turn, was built from protons and neutrons. But with the advent of particle accelerators producing energies of the order of 100 MeV and the development of cosmic ray research, the hitherto "elementary" particles like nucleons began to exhibit internal structure. The number of detected particles was growing rapidly.

By 1947 the picture of the world consisted of 16 particles: **nucleons**: neutrons and protons (and their respective anti-particles), **photons**, **leptons**: electrons, muons, electron- and muon- neutrinos (and their anti-particles) and, finally, 3 π-mesons: π^+, π^0 and $\overline{\pi}$. The picture seemed quite clear and the role of each particle understandable until, at the end of the year, Rochester and Butler, using a cloud chamber, detected tracks of new particles in cosmic rays. These particles are now known as meson K^0 and M particle. Two years later Powell and his associates detected another particle and called it a τ^+-meson[9]. The newly discovered particles were given the common name "strange", as not fitting the hitherto accepted picture of the world. Their number grew rapidly after accelerators providing energies of the order of GeV were constructed in the early 1950s.

The "strange" aspect of all these particles became evident upon comparing their creation and decay times: they were apparently created in strong interactions but their long lifetime indicated that they could only decay in weak interactions. This led to the postulation of a new quantum number – strangeness (S) – that in addition to charge, energy, baryon number, lepton number and isospin, was claimed to be conserved in strong interactions.

In 1953 Gell-Mann and Nishijima independently formulated their famous law that relate the quantum numbers mentioned above for any elementary particle:

---

9 See e.g. (Norwood 1975).

$$Q = I_3 + \frac{B+S}{2}$$

where: Q is the charge of a particle in units of e, I_3 is the third component of isospin, B is the baryon number and S is the value of strangeness.

The Gell-Mann–Nishijima formula was soon found to hold also for newly discovered baryon and meson resonances. The number of known particles, however, soon exceeded 100 and physicists began to look for an ordered pattern in this perplexing plethora of "elementary" particles. Some have even expressed their doubts as to the "elementary" character of known particles believing that they had to be made up of simpler components.

Already in 1950 Eugene Wigner[10] showed that group theory, a rather abstract branch of mathematics, proved very useful in theoretical physics, namely quantum mechanics of atomic spectra. The group-theoretical formalism was soon extended to the analysis of certain interactions in nuclear physics and elementary particle physics[11]. To put it in general terms, the usefulness of group theory lay in the fact that the basic quantity describing quantum-mechanical system – the interaction Hamiltonian – was invariant under certain transformations (like translations in space or time, rotations and so called gauge transformations). This, according to the Noether theorem, implied that certain quantum numbers describing the system (eg. energy, momentum, charge, spin or isospin), that corresponded to the group generators, were conserved.

Perhaps the first successful application of these group-theoretical concepts to elementary particle physics was connected with the concept of *isospin*. Analysis of the properties of atomic nuclei suggested that the proton and neutron may in fact be treated as two different states of the same particle – *nucleon*, which possessed an intrinsic property taking one of the two possible values. This property was called *isospin*, or *isotopic spin* by analogy to the quantum number of spin whose third component could also assume one of the two possible values: +½ and –½[12]. Accordingly, the third component of isospin (I_3)

10 See his classic work (1958).

11 For a formal, exhaustive exposition of group theory and its applications to microphysics see (Hammermesh 1963).

12 See, for example, (Norwood 1975).

equals +½ for proton and −½ for neutron. The concept of isospin, originally formulated for the nucleon, was subsequently generalized to other groups of strongly interacting particles (hadrons) of very similar properties. All the particles in a group (called an isospin multiplet) had nearly identical quantum numbers and differed only in charge, connected with electromagnetic interactions. The mathematical formalism of isospin is formally identical to the well known spin formalism, and was based on the so-called *SU(2)* symmetry group.

By 1963 some of the particles had been grouped into larger families according to their charge, isospin, baryon number and strangeness. These families exceeded the scope of isospin symmetry and were implied by a broader symmetry generated by isospin and strangeness. One such family called the baryon octet is composed of two isospin doublets (n and p; Ξ and Ξ^0), one singlet (Λ) and one triplet (Λ, Σ^0, Σ^+) of different strangeness. All particles in the baryon octet have very similar masses and spin^{parity} equal to $½^+$.

The structure of the baryon octet is shown in Figure 1:

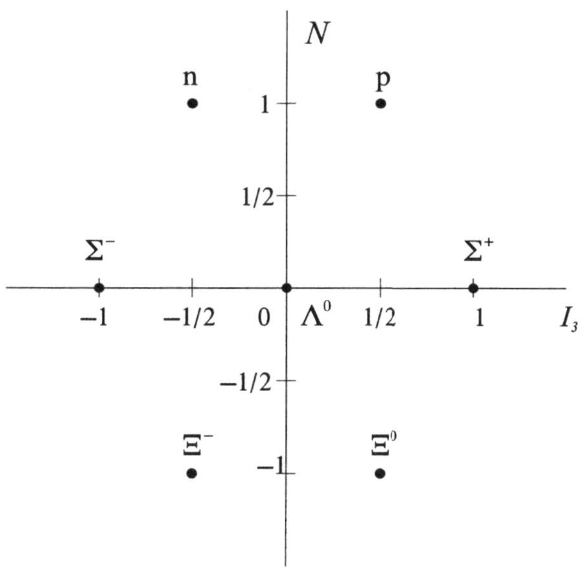

Figure 1: The structure of the baryon octet.
The symmetry is readily apparent in the plot.

A very similar structure is constituted by 8 pseudo-scalar mesons: pions, kaons, anti-kaons, and the η-meson. These two structures appear identical if, instead of strangeness *S*, we employ hypercharge *Y* to describe the multiplets where $Y=B+S$. Other families are: 9 vector mesons (1^-), 9 vector mesons (2^+), and the baryon resonance decuplet ($3/2^+$)[13].

The new "symmetry" exhibited by families of particles was less "strict" than isospin symmetry – differences in masses of particles were greater than in isospin multiplets. According to a generally accepted insight concerning symmetry braking of stronger interactions by weaker ones, this indicated that some super-strong interactions were involved, whose symmetry was broader than the mere isospin symmetry of strong interactions. A closer analysis led to the conclusion that the symmetry group appropriate to describe the above families was $SU(3)$ (a three-dimensional unitary symmetry group) one of whose generators (a counting operator N) is shown in Figure 1.

The first unsuccessful attempt to incorporate hadron families into $SU(3)$ multiplets was Sakata's model of particle algebra. He chose a neutron, a proton, and an Λ-particle as the basic elements (called *sakatons*) from which he constructed the $SU(3)$ group algebra (Halzen, 1984). Mesons were postulated to consist of a sakaton–anti-sakaton pair, and baryons of two sakatons and one anti-sakaton. Although Sakata's model correctly predicted the structure of meson multiplets, it completely failed with baryon families.

In 1964 Gell-Mann and Ne'eman found the correct representations for all known multiplets within the $SU(3)$ group. They introduced into the theory entirely new particles of fractional charge to occupy the basic representation of that group. These particles were called *quarks*. The decomposition of $SU(3)$ representations into Clebsch-Gordan series of irreducible representations (Gell-Mann and Ne'eman, 1964) led to the conclusion that baryons consisted of three quarks and mesons of a quark – anti-quark pair.

Initially three quarks (called *up*, *down*, and *strange*) were sufficient to explain properties of hadron families. Their quantum numbers (called *flavours*) could easily be determined on theoretical grounds so as to fit the hypercharge-strangeness structure of the basic representation of $SU(3)$ and, at the same time, explain the properties of hadrons. Soon, however, the fourth quark (called *charmed*) was postulated to fit the expected symmetry between quarks and leptons. Now physicists believe that there are still two more quarks (*charmed* and *true*) so that the total number of quarks (and their quantum numbers – flavours) equals 6 (Białkowski 1971; Halzen 1984). Their properties are shown in Table 1:

---

13 For an excellent exposition of elementary particle symmetries and their quark structures see (Gell-Mann and Ne'eman 1964).

quarks	B	I_3	S	Q	C	B'	T
u	1/3	1/2	0	2/3	0	0	0
d	1/3	−1/2	0	−1/3	0	0	0
s	1/3	0	−1	−1/3	0	0	0
c	1/3	0	0	2/3	1	0	0
b	1/3	0	0	−1/3	0	−1	0
t	1/3	0	0	2/3	0	0	1

Table 1. Quarks and their quantum numbers (flavours). B – baryon number, I_3 – third component of isospin, S – strangeness, Q – charge, C – charm, B' – beauty, T – truth.

To sum up considerations of this section: Quarks emerged in an attempt to understand, make sense of, and simplify a variety of "elementary" particles that after World War II rapidly started growing in number. The choice of an appropriate quark model, including the quark constitution of various hadrons and quantum numbers of quarks, was strongly motivated by mathematical formalism of quantum field theory. The criteria employed in comparing various possible quark models could be based on intrinsic mathematical coherence of theory. The criterion of coherence with observations (in a sufficiently broad sense of the word), that is with properties of observed particles, would play a relatively minor role.

3. How GELL-MANN Proceeds to Discover Quarks?

In this section I will trace the way in which the computer system GELL-MANN rediscovers quarks, and show that although it arrives at exactly the unique quark model accepted in physics, it proceeds in a radically different manner and employs different criteria of theory choice than did the physicists.

GELL-MANN is a Common LISP system. Its task was to analyze elementary particle data available in 1964 and formulate a hypothesis (or hypotheses) about a "hidden", more simple structure of matter, or – to put it in contemporary terms – to discover quarks. However, instead of applying sophisticated theories of relativistic quantum fields and group-theoretical formalism, it performs an exhaustive search in a vast space of possible quark models and compares each model with "observable" data concerning elementary particles. In what follows I shall briefly describe the system's input, output, functioning, operators, heuristics, and the results it obtained.

Input. As input GELL-MANN requires a list of objects whose attributes assume rational values. All the input objects must be characterized by the same attributes. The intended input on which the system is supposed to succeed is, of course, the "symmetric" families of elementary particles known in 1964. The input is limited to one family at a time and GELL-MANN can use the quark models derived in the previous run and try to augment them to fit the new input.

Output. The output of the program consists of a list of quark models. Each model consists of a list of quarks and their properties and a unique quark composition for each input particle.

The processing. GELL-MANN reasons in several steps. First, it postulates the number N of types of the hidden objects (quarks), then the number M of components in each elementary particle in the input family. Then, if the number of possible quark combinations is compatible with the number of input particles, the program starts working on the quark constitution of each particle and the values of each property for each quark. At this stage simplifying heuristics are necessary to cut down on computational complexity. GELL-MANN performs an exhaustive search for consistent quark models until it reaches N and M for which the first quark model consistent with the input has been found.

Operators. The system works by subsequent application of a number of operators each of which makes a specific contribution to the construction of a quark model.

O1: Postulating micro types. This operator postulates that all particles in the input are made of quarks of N types If the program has no quark model from the previous run, it starts with $N=1$ and increases it until $N=K$, where K is the number of particles in the input. This is to rule out models in which the number of quarks is greater than the number of particles in the input.

O2: Postulating micro structure. The system postulates that each particle consists of M quarks starting with the simplest nontrivial hypothesis, for $M=2$, then considers groups of three quarks, four quarks, and so on, for all values of M such that the number of combinations of N quarks in groups of M, $C(N,M)$ is less than $2K$ (see heuristics below), This operator enforces homogeneity of structure allowing all particles in a family to contain only the same numbers of quarks. Heuristics do not allow $M=1$ for in this case quarks would be identical to the particles.

O3: Postulating possible structures. This operator generates all combinations $C(N,M)$ of N quarks in groups of M. It requires the choices made by O1 and O3 and produces data used in termination of O2.

O4: Postulating micro values. This operator performs its search in a space based on the Cartesian product of three lists: quarks, attributes, and the possible values for each attribute. The list of quark names is generated by O1, the list of attributes is limited to all and only the attributes for the input particles in order to obtain observation feedback. For each attribute the scope of values is determined by the values in the input particles and the values of quarks already known to the system. For each attribute GELL-MANN finds its highest absolute value V and all denominators of fractional values for the input particles. Then it generates a list of all integers and fractions based on collected denominators between $-V$ and V. When the program cannot find a solution for these values, it considers the fractions based on $1/M$, where M is the number of quarks in a particle. The number of possible assignments of one property with L values to N quarks is L to the N-th power. For several properties it is enormous, which makes straightforward analysis practically impossible. For this reason the program needs some simplifying heuristics to decompose the problem into independent subproblems.

O5: Assigning structures to particles. This operator maps each input particle on a different quark combination generated by O3.

GELL-MANN applies its operators in a certain order: first O1 performs its search, followed by O2, and then by O3. O3 can, in principle, operate independently of O1 and O2, while the search generated by O5 is coupled with O4 and must follow O3.

Heuristics. Operators of the system use several heuristics to simplify the search by problem decomposition and by failing inconsistent partial solutions. Some of these heuristics are fairly *ad hoc* but, as the program's authors admit, it could not do without them. Nevertheless, GELL-MANN performs an exhaustive search in that it does not skip any consistent solutions. Below I will briefly describe these heuristics.

H1: Unique decomposition. Any quark model must contain at least as many quark combinations as there are particles in the input family in order for each particle to have unique quark decomposition. Excluded are those models for which $C(M,N) < K$.

H2: Matching the family. On the other hand, the number of quark combinations cannot be too large for this would create the potential for too many quark models. Ideally, the number of quark combinations would be equal to the number of particles in the input, even the weaker condition described in O2 suffices to limit the search and obtain a unique solution. H1 and H2 are applied to terminate some partial solutions generated by O3.

H3: Search decomposition. A straightforward search conducted by operators O4 and O5 would cover all cases but is practially unfeasible because of computational complexity. For this reason the system considers each attribute separately, finds all solutions for that attribute, and then tries to unify partial solutions. The actual search is guided by heuristics H4, H5, H6, H7 and H8, while the straightforward search would only need the Additivity Law implied by heuristic H4.

H4: Additivity Law. The value of each attribute for a compound particle is calculated as the sum of respective values of the component quarks. This heuristic applies after O4 generates the values for all quarks in a given combination and after O5 maps that quark combination to a particle. Solutions that violate this law are rejected.

H5: Lexical combinations. For any given attribute the system considers a list of all lexically ordered combinations of the length N of values of that attribute. The list is much smaller than the list of all possible value assignments to N quarks described in O4.

H6: Frequency test. Only those partial solutions are accepted which pass the frequency test: for any given attribute, the frequency distribution for a given set of quark combinations must equal or exceed that for the input family of particles. Only then, according to the authors, can each particle be explained by a unique combination of quarks. This heuristic utilizes global information about an input family and cuts off many branches in the search tree.

H7: Base attribute. GELL-MANN considers first the attribute with the highest number of permutations of values of attributes for solutions that have passed the frequency test. This is to provide the most effective reduction of search.

H8: Unification of attributes. The system tries to unify partial solutions by considering the Cartesian product of values' distributions among quarks for all attributes. Then heuristic H4 rejects most of the solutions as globally inconsistent.

H9: Elimination of inconsistent models. Working in the incremental mode GELL-MANN processes one particle family after another considering all previous models, and returns only those which explain the new input. In this process old models may be augmented with new quarks. This heuristic dramatically reduces the search time.

Results. GELL-MANN can process input families in two ways. In the first mode each family is considered separately with no knowledge of the previous runs. In the second mode, the system considers inputs incrementally, augmenting the models from one input to another and discarding those models which are inconsistent for some inputs. Fisher and Żytkow tested GELL-MANN on three families of particles I described in Section 2: the baryon octet, the baryon resonance decuplet, and the meson octet. The program stops at the first N and M for which at least one quark model consistent with the data is found. However, the authors admit that ". . . not always finding a model indicates success. If GELL-MANN . . . finds many models, or if it halts without finding any, it is said to fail. . . . If many hypotheses of the same complexity are equally consistent with data, we do not have a reason to select one because we cannot tell that any of them corresponds to reality better than any other." (Fisher and Żytkow 1990, p. 367). I will return to this important issue connected with criteria of theory comparison in the next section.

For the baryon octet the program found a unique quark model for three quarks in groups of three. This solution corresponds exactly to the quark model developed by the physicist Gell-Mann, and it turned out to be the only adequate model in its simplicity class. If the system was allowed to generate fractional values for quark attributes (operator O4), it found one more solution for three quarks in groups of three, but with different values of quarks attributes. Nevertheless, the authors consider this other model to be more complex than the one accepted in physics, because of its fractional values for strangeness.

For the baryon resonance decuplet, GELL-MANN found a single solution for three quarks in groups of three, identical to that for the hadron octet. It is exactly the model accepted in physics.

The situation is much more difficult for GELL-MANN with the meson octet. Here the model accepted in physics consists of six quarks (three quarks and three anti-quarks) with mesons being made up of quark – anti-quark pairs. First, the program arrived at one incorrect solution with four quarks (two quarks and two anti-quarks). Then, with 5 quarks in groups of two it arrived at 997 solutions. Finally, it was applied on combinations of 6 quarks in groups of two, but, unfortunately, the authors could not complete the computations after two days of CPU time on a fast UNIX workstation.

It was doubtful whether a unique solution could be found. With so many hypotheses we cannot say that the model accepted in physics was *discovered*, because it was one of many models of the same complexity, generated.

The situation radically changes with the incremental run. Here GELL-MANN considers only those models which are consistent with the input it has seen so far. First, the system was run on the baryon octet and, again, found two models consistent with the data. Then, analyzing the baryon decuplet, it realized that only one model can be successfully applied to the new family of particles, that is the model accepted in physics. Finally, when given the meson octet as the next input, the program tried to apply to it the model with three quarks, but it failed. Then it augmented the model by adding the fourth, fifth, and, eventually, the sixth quark. After evaluating all hypotheses concerning values of attributes for three new quarks, GELL-MANN arrived at precisely the correct model accepted in physics and returned it as *the only* model consistent with the data.

4. Conclusion

In the previous section I described a computer program which, by performing an exhaustive search through a vast space of possible quark models, found only one solution consistent with the data – precisely the model accepted in physics. Fisher and Żytkow go even further and say that "This shows that Gell-Mann's model stands out so distinctly that it can be validated solely on the simplicity and uniqueness criteria, without any further arguments based on sophisticated theories . . ." (Fisher and

Żytkow 1990, p. 370). The situation would, of course, be different if the system produced many competing models so that other theories would be necessary to compare them and select among them. This reveals an important difference between the criteria of theory comparison used by theoretical physicists and computer systems. These two groups of criteria function, so to speak, at different levels of theoreticity.

Nancy Cartwright, in her famous book *How the Laws of Physics Lie?* (1983) introduced an important distinction into *theories* and *phenomenological laws* to replace the traditional theory-observation dichotomy which, for her, is irrelevant to science. The function of high-level scientific theories (like quantum field theory) is to *explain* broad ranges of phenomena, whereas phenomenological laws (like the exponential radioactive decay law) *describe* specific phenomena. It is at the level of these very well confirmed laws, if anywhere, that truth in physics can be located, says Cartwright. Now, putting asidc Cartwright's interesting anti-realism with respect to theoretical laws and her causal-entity-realism[14], the distinction she introduced is very well motivated and, I believe, provides a good framework for our discussion.

Considerations that led theoretical physicists to the appropriate quark model were carried out at the level of theoretical models and group theory formalism of quantum field theory. Most of the possible quark models considered by the computer system GELL-MANN would be *a priori* excluded as not fitting the representations of the $SU(3)$ symmetry group. The major criterion employed here to compare various quark models was connected with the intrinsic, mathematical coherence of the formalism. After this criterion had been used, the choice of appropriate values of quark attributes and the quark constitution of hadrons was a relatively unproblematic task.

On the other hand, no serious contemporary scientist would proceed the way the program GELL-MANN actually did. Since Democritus, philosophers speculated about the hidden atomic structure, but they could not justify any concrete properties of atoms. It is easy to postulate hidden structure but it is very difficult to justify it, because so many models are possible. This problem is known to contemporary philosophers of science as *underdetermination of theory by data.* A human scientist, could not even hope to justify any proposed quark model at the level of phenomenological laws alone, that is without resorting to sophisticated theoretical considerations. To do that, he would have

---

14 I discuss the relevance of Cartwright's view of physics to the debate over realism in science in (Giza 1987).

to show not only that his model is consistent with the data, but also that it is the simplest one. But this requires the consideration a great many models (in the case discussed, it is several hundred thousands) to use the simplicity criterion, that is to show that no equally simple or simpler model exists which is also consistent with the data. Nevertheless, the machine discoverer, GELL-MANN, proceeds exactly this way and this should occasion no surprise.

The fact that GELL-MANN proceeds as only an insane scientist would could be generalized to all contemporary machine discovery systems. It is obvious to anyone familiar with computers and programs that what is difficult for humans is easy for computers, and *vice versa*. For example, computers can analyze vast amounts of information and perform complicated calculations in seconds, while humans would need months, or years to complete the same task and would probably give it up.

On the other hand, computer systems, at least at the present stage of computer science, find it very difficult to do many things that we, humans, find natural and easy – like smiling, playing jokes, or watching the world around us. And, in any case, they cannot learn relativistic quantum field theory and work in theoretical physics departments.

Even if criteria used by computer discoverers, based on simplicity, uniqueness, and consistency with the data, cannot fully justify the results they achieve, it is nevertheless a significant step forward that theory-poor physics discoveries can be made which coincide with results obtained by means of sophisticated, theoretical reasoning.

Department of Philosophy
Maria Curie-Skłodowska University
M. Curie-Skłodowskiej 4
20-031 Lublin
Poland

REFERENCES

Białkowski, G., Sosnowski, R. (1971). *Czastki elementarne* [Elementary Particles]. Warszawa: PWN.

Bradshaw, G.L., Langley, P. and Simon, H. (1980). BACON.4: The discovery of intrinsic properties. In: *Proceedings of the Third Biennial Conference of the Canadian Society for Computational Studies of Intelligence*. Victoria: BC, pp. 19-25.

Cartwright, N. (1983). *How the Laws of Physics Lie?* Oxford: Clarendon.

Fisher, P., Żytkow, J.M. (1990). Discovering Quarks and Hidden Structure. In: Z. Ras, M. Zemankova and M.L. Emrich (Eds.), *Methodologies for Intelligent Systems 5*. New York: Elsevier Science Publishing Co., pp. 362-370.

Gell-Mann, M. and Ne'eman, Y. (1964). *The Eightfold Way*. New York: Benjamin.

Gerwin, D.G. (1974). Information Processing, Data Inferences, and Scientific Generalization. *Behavioral Sciences* **19**, 314-325.

Giza, P. (1987). How to Defend Realism? In: *Proceedings of the 11th International Wittgenstein Symposium*. P. Weingartner, G. Schurz (Eds.), *Recent Development in Epistemology and Philosophy of Science* **14**. Wiena: Hölder-Pichler-Tempsky, pp. 39-43.

Halzen, F. (1984). *Quarks and Leptons: An Introductory Course in Modern Particle Physics*. New York: Wiley.

Hammermesh, M. (1963). *Group Theory*. NY: Addison-Wesley.

Langley, P. and Żytkow, J.M. (1989). Data-Driven Approaches to Empirical Discovery. *Artificial Intelligence* **40**, 283-312.

Langley, P., Simon, H., Bradshaw, G.L. and Żytkow, J.M. (1987). *Scientific Discovery: Computational Explorations of the Creative Processes*. MA: MIT Press.

Nordhausen, B. and Langley, P. (1993). An Integrated Framework for Empirical Discovery. *Machine Learning* **12**, 17-47.

Norwood, J. (1975). *Twentieth Century Physics*. NY: Prentice Hall.

Rose, D. (1989). Using Domain Knowledge to Aid Scientific Theory Revision. *Proceedings of the Sixth International Workshop on Machine Learning*. San Mateo, CA: Morgan Kaufmann Publishers, pp. 272-277.

Wigner, E. (1958). *Group Theory and its Application to Quantum Mechanics of Atomic Spectra*. New York: Academic Press.

Żytkow, J.M. (1987). Combining many searches in the FAHRENHEIT discovery system. In: *Proceedings of the Fourth International Workshop on Machine Learning*. Irvine CA, pp. 281-287.

Poznań Studies in the Philosophy
of the Sciences and the Humanities
2000, Vol. 72, pp. 105-132

Henryk Ogryzko-Wiewiórowski

METHODS OF SOCIAL CHOICE OF SCIENTIFIC THEORIES

ABSTRACT. The criteria for the evaluation of scientific theories are considered to be one of the most important problems in the contemporary philosophy of science. However, there is no agreement as to which criteria of evaluation are most essential. Scientists differ in their opinions with respect to the importance of particular properties of theories. In this paper the author considers the problem of evaluating scientific theories on the basis of the social choice theory. He presents a matrix-hierarchical conception of social choice with three models: a model of one-dimensional choice, a model of multidimensional social choice with fixed criteria of individual preferences and a model of multidimensional social choice with varying criteria of individual preferences. Such a formulation of the problem at hand makes it possible to evaluate scientific theories in the light of any number of criteria.

Contemporary philosophy of science investigates the problem of choosing a scientific theory. Scholars are striving to establish rational criteria of choice between competing theories. Endogenous methodological criteria fail. In one of his works S. Amsterdamski puts forward the following question.

> Is methodology able to explain the process of replacing some theories by others as a result of applying definite, universally valid rules of conducting research, rules that can be codified and exposed rationally? (Amsterdamski 1983, pp. 184-185).

After further analysis, Amsterdamski gives a negative answer. Methodology cannot supply such criteria. Does this entail complete liberty in the issue? Not necessarily. We know that in spite of philosophers' doubts, science does make progress; scientists achieve more and more success and make more discoveries. This means that research communities have some criteria of evaluating scientific theories, at least in the long run. Therefore, as a rule, better and more valuable (from the viewpoint of the cognitive process) theories survive. These criteria are derived from the more or less commonly accepted (methodological?, cognitive?) values which a good scientific theory ought to conform to. Some of these values are: simplicity (of a theory), generality, informative content, empirical content accuracy, coherency, explanatory power, unification, predictive power, technological power, etc. This set of values is neither exhaustive nor commonly accepted. Researchers have different opinions with respect to the importance of particular elements of this set, as well as to its content. Nevertheless, even in the situation of such divergence of opinions, with respect to the accepted values or to their significance to the process of scientific investigation, it is possible to decide which of the existing theories fulfils the requirements in the most satisfactory way, according to accepted scientific and methodological procedures. To this end, terminology of the so-called theory of social choice turns out to be useful.

Foundations of the theory of social choice

The theory of social choice dates back to the 18th century and was derived from research on problems involving political elections; I am referring to thinkers such as Jean-Charles de Borda (1733-1799)[1] or Jean Antoine Nicolas Marquis de Condorcet (1743-1794)[2]. The latter proved that the generally employed method of simple majority sometimes leads to no resolution and to a so-called voting paradox. Let us assume that three people (1, 2, 3) make their choices from three alternatives (*x*, *y*, *z*). Let each person establish the following order of preference on the set of alternatives.

1	2	3
x	*y*	*z*
y	*z*	*x*
z	*x*	*y*

Applying the rule of majority results in a draught. There appears to be a loop in social preferences. The majority (2 people) prefer *x* over *y*, and the same number of people prefer *y* over *z* but the majority prefer *z* over *x*. Marquis de Condorcet conducted a lot of research to avoid the voting paradox, however, it did not lead to satisfactory results. In the 19th century there was a search for a compromise solution. Many theories of social choice were formulated but each of them contained serious flaws[3]. As a result of that research, the theory of social

[1] (Borda 1781). English translation (Grazia 1953).

[2] (Condorcet 1785).

[3] I treat this subject more fully in my (1986).

choice was abandoned for some time. Kenneth J. Arraw took it up again in the middle of the 20th century. In his famous book *Social Choice and Individual Values*[4]. Arrow proved the existence of a certain impossibility. He formulated the thesis that it will never be possible to construct a method of group choice (one that generates social ordering of alternatives) that can fullfil the five basic formal conditions.

1. *The Social Welfare Function* ought to be defined on arbitrary profiles of individual orderings, i.e. independent of the configuration in which alternatives occur in preference to the people making their choices.

$$R_1, R_2, \ldots, R_n$$

Arrow also requires that the set of alternatives should be not smaller than 3 ($S \geq 3$) and that the number of voters should be at least 2 ($N \geq 2$).

$$S \geq 3 \quad N \geq 2$$

The second condition involves *the positive association of social and individual values.*

2. The method of social choice ought to be sufficiently sensitive to changes in the positions of alternatives in individual orderings. Let $R_1,...,R_n$ and $R'_1,...,R'_n$ be two sets of weak relations on profiles of individual preferences, and R and R' be respective social orderings, and P' be strong relations of social preferences. Let us assume that for every voter i ($i = 1,..., n$) two relations of individual orderings are associated in the following way: for any two alternatives x', y' different from some alternative x, $x'R'\,y'$ obtains iff $x'R_i\,y'$ for any y', $xR_i\,y'$ implies $xR'_i\,y$; moreover for any y', $xP_i\,y'$ implies $xP'_i y'$. Then xPy implies $xP'y$. To express the same in more intuitive language, we can conclude that if in two sets of individual preferences relations of any two alternatives are similar, then relations between these alternatives in according social ordering will also be convergent.

3. Social ordering resulting from the application of a given social welfare function (a method of social preference) ought to depend only

---

4 (1951).

on relations obtained from the set of possible (available) alternatives, S and not on the context in which the set (relations an this set) occur (*The Independence of Irrelevant Alternatives*). Let $R_1,...,R_n$ and $R'_1,...,R'_n$ be two individual preference profiles, and R and R' accordingly, relations of social preferences. Then, if for all $i \in N$ (the set of individuals, i.e. people making their choices) and $x, y \in S$ the equivalence: $xR_i y \Leftrightarrow xR'_i y$ is fulfilled, then R and R' constitute identical orderings of alternatives on the set S. (In his (1951) book, instead of talking about relations of social preferences R and R' Arrow discussed sets of preferred alternatives $C(S)$ and $C'(S)$. Using these terms, however, he stresses alternatives occupying the highest place(s) in social orderings but the fundamental logical sense is the same in both cases).

The fourth condition, called the citizens' sovereignty, states that the result of group choice ought to be determined only by the preferences of the members of a given community. When this condition is not fulfilled a given social welfare function is dictated, i.e. it disregards the wishes of the voters, at least for some alternatives.

DEF 1. A social welfare function is dictated if for any pair of alternatives $x, y \in S$ it gives the social ordering xRy regardless of the profile of individual orderings $R_1,...,R_n$ that correspond to it.

Therefore, this forth condition is as follows:

4. The social welfare function cannot be dictated.

This last conditions protects social choice from the dominance of one voter over the others (the condition of nondictatorship). A function of social choice that does allow domination is called *dictatorial.*

DEF 2. A social welfare function is dictatorial if there exists a voter i ($i \in N$) such that for all $x, y \in S$, xPy implies social orderings xPy regardless of the orderings of the other voters:

$$R_1, \ldots, R_{i-1}, \ldots, R_n$$

Thus, Arrow requires that:

5. The social welfare function cannot be dictatorial.

The above conditions are there to guarantee democracy of social choice and to make its result dependent only on the will of the majority of voters. From a moral point of view each of them is justified. The problem is that from a logical point of view they cannot be obtained simultaneously. As Arrow demonstrated, there is no social welfare function that would fulfil all of the five conditions. It is stated by the famous impossibility theorem.

Theorem 1 (the impossibility theorem)

If $S \geq 3$ and $N \geq 2$, then there is no social welfare function (generating a weak ordering on the set of alternatives S) that would fulfil conditions **1-5**; each social welfare function fulfilling conditions **2** and **3** is in fact either dictated (i.e. violates condition **4**) or dictatorial (violates condition **5**)[5].

The impossibility theorem, published by Arrow in 1951, spurred a wave of discussions in academic circles dealing with the problem of social choice. Arrow's results especially aroused those in economic circles because it undermined the basis of economics of welfare, which had been under intense development for many years. Many critical works appeared. There emerged a whole spectrum of opinions, from those that challenged the credibility of Arrow's thesis to those that questioned the existence of the so-called public interest (whose voice social ordering was said to be). After the first years of intellectual shock, the following decades resulted in more balanced texts that mainly aimed to weaken the conditions[6]. Hundreds of papers and many books were published on the subject. As time went on, a certain compromise solution was reached[7]. It turned out that the basic conditions of social choice can be made coherent only if transitivity of the base relation is limited to its closure. And the rule that can fulfil such a limitation is the rule of simple majority based on pairwise comparison of alternatives[8].

The remaining part of this paper shall concentrate on extending the rule of majority for multidimensional choice.

---

5 (Arrow 1951, p. 59).

6 See (Ogryzko-Wiewiórowski 1986).

7 See (Richelson 1978).

8 (Bordes 1975). Cf. also (Ogryzko-Wiewiórowski 1986, pp. 110*ff*).

The matrix-hierarchical conception of social choice

The essence of social choice performed by using the method of simple majority is the priority of alternatives that are most often preferred over the other comparable elements of the set S

$$xMy \Leftrightarrow \{|i|: x >_i y\} \geq \{|i|: y >_i x\}$$

where M (majority) denotes a relation generated by the rule of majority on the set of available alternatives (subject to choice) S. By this we mean that the alternative x precedes, in the sense of the relation M, the alternative y if at least half of the number of voters prefer x over y. When such a situation occurs, with respect to every pair of alternatives, our result is a group (social) ordering of elements of the set S. We can therefore conclude that each collective ordering of alternatives $x_1, x_2, ..., x_n$ has a coordinated scale: $\omega_1, \omega_2, ..., \omega_n$, whose values are indicators of social affirmation of respective alternatives. There arises, then, a problem of finding a mathematical model that could use the basis of individual choices, based on comparing pairwise comparison of alternatives on the scale of values $\omega_1, \omega_2, ..., \omega_n$ reflecting affirmation levels of alternatives from the set $S = \{x_1, x_2, ..., x_n\}$.

At this point, it is useful to refer to research done by T.L. Saaty and J.P. Bennett[9] and also to the earlier works of B.G. Mirkin[10]. In these works, we came across an interesting idea of applying eigenvalues and eigenvectors (of matrices) to determine social ranking of the compared alternatives. The authors then considered the matrix of comparisons $A = [a_{ij}]$, whose elements possess the following properties:[11]

1°
$$a_{ij} = \frac{w_i}{w_j}$$

Contrary to what the authors of both works assume, it suffices to assume that the values w_i and w_j are greater than and positive. Consequently, from the property 1° there follows other properties of the matrix[12] A.

---

9 (Saaty and Bennett 1977).

10 (Mirkin 1974).

11 (Saaty and Bennett 1977, p. 238); (Mirkin 1974, p. 183).

12 (Saaty and Bennett 1977, pp. 238-39). In (Mirkin 1974, p. 182) is $a_{ij} = 1 - a_{ji}$.

2°
$$a_{ij} > 0$$

3°
$$a_{ij} = \frac{1}{a_{ij}}$$

4°
$$a_{ii} = 1$$

5°
$$Tr(A) = \sum_{i=1}^{n} a_{ii} = n$$

A fulfills the consistency condition if

6°
$$a_{ij} a_{jk} = a_{ik}$$

The following property requires proof (supplied neither by Saaty and Bennetty nor by Mirkin).

Theorem 2

7°
$$rz\ A = 1$$

The matrix A can be represented as a product of the following two matrices B and C

$$A = BC = \begin{bmatrix} w_1 & . & . & . & . & . \\ . & w_2 & . & . & . & . \\ . & . & . & . & . & . \\ . & . & . & . & . & . \\ . & . & . & . & . & . \\ . & . & . & . & . & w_n \end{bmatrix} \begin{bmatrix} \frac{1}{w_1} & \frac{1}{w_2} & . & . & . & \frac{1}{w_n} \\ \frac{1}{w_1} & \frac{1}{w_2} & . & . & . & \frac{1}{w_n} \\ . & . & . & . & . & . \\ . & . & . & . & . & . \\ \vdots & \vdots & . & . & . & \vdots \\ \frac{1}{w_1} & \frac{1}{w_2} & . & . & . & \frac{1}{w_n} \end{bmatrix}$$

We know that B is a nonsingular square matrix of the degree n. It is however easy to notice that C contains only one linearly independent row. Then,

$$rz\, C = 1$$

In this situation, the rank of the matrix A is determined by means of the following theorem:

Theorem 3

If B, C are square matrices of the dimension $n \times n$ and B is a nonsingular matrix, then[13]

$$rz\, BC = rz\, C$$

Because B is nonsingular, the rank of product BC, or the rank of A is equal to the rank of C, i.e.

$$rz\, A = rz\, C = 1$$

QED.

The basis of further properties of A shall be Perron's theorem.

Theorem 4 (Perron's)

The real matrix A of elements $a_{ij} > 0$ always has a real and positive eigenvalue λ, which has the following properties:

1′ it is a single root of a characteristic equation:

2′ it is the greatest eigenvalue of A and of all components of the eigenvector with respect to the absolute value

$$x = col(x_1, x_2, \ldots, x_n)$$

corresponding to this, values are positive[14].

---

13 (Opial 1976, p. 346).

14 (Kowalczyk 1976, p. 100).

These properties of A let us formulate a theorem that expresses more precisely the problem of its eigenvalues.

Theorem 5

The matrix A defined by properties 1° – 7° has one eigenvalue, equal to its trace:

$$\lambda_1 = n = Tr(A)$$

and the (n–1)-fold eigenvalue equal zero:

$$\lambda_2 = \lambda_3 = \ldots = \lambda_n = 0$$

Proof:

Eigenvalues are zeroes of the determinant[15]

$$\det(A - \lambda I) = \begin{vmatrix} a_{11} - \lambda & a_{12} & \ldots & a_{1n} \\ a_{21} & a_{22} - \lambda & \ldots & a_{2n} \\ \cdot & \cdot & \cdot & \cdot \\ a_{n1} & a_{n2} & \ldots & a_{nn} - \lambda \end{vmatrix}$$

where I is the unit matrix

$$\det(A - \lambda I) = \begin{vmatrix} (1-\lambda)\frac{w_1}{w_1} & \frac{w_1}{w_2} & \ldots & \frac{w_1}{w_n} \\ \frac{w_2}{w_1} & (1-\lambda)\frac{w_2}{w_1} & \ldots & \frac{w_2}{w_n} \\ \cdot & \cdot & \cdot & \cdot \\ \frac{w_n}{w_1} & \frac{w_n}{w_2} & \ldots & (1-\lambda)\frac{w_n}{w_n} \end{vmatrix} =$$

15 Ibid, p. 88.

$$= (w_1, w_2, \ldots, w_{n)} \begin{vmatrix} \frac{1-\lambda}{w_1} & \frac{1}{w_2} & \cdots & \frac{1}{w_n} \\ \frac{1}{w_1} & \frac{1-\lambda}{w_2} & \cdots & \frac{1}{w_n} \\ \frac{1}{w_1} & \frac{1}{w_2} & \cdots & \frac{1-\lambda}{w_n} \end{vmatrix} =$$

$$= \begin{vmatrix} 1-\lambda & 1 & \ldots & 1 \\ 1 & 1-\lambda & \ldots & 1 \\ . & . & . & . \\ 1 & 1 & \ldots & 1-\lambda \end{vmatrix} = \lambda^{n-1}(\lambda - n)(-1)^n$$

Conclusion:

$$\det(A - \lambda I) = \lambda^{n-1}(\lambda - n)(-1)^n$$

which means that eigenvalues are zeroes of the polynomial:

$$p(\lambda) = \lambda^{n-1}(\lambda - n)(-1)^n$$

This polynomial equals zero if $\lambda = \text{n}$ and if $\lambda = 0$, where $\lambda = \text{n}$ is a single root of the characteristic polynomial $p(\lambda)$ and $\lambda = 0$ is (n–1) fold root of this polynomial. Because eigenvalues of A are by definition roots of the characteristic polynomial, so $\lambda = 0$ is the $(n–1)$-fold eigenvalue and $\lambda = n$ is the single value of A.

QED.

From the second part of Perron's theorem it follows that all coordinates of the eigenvector associated with the greatest eigenvalue of A are positive. This is essential to our considerations for from the fact that the matrix A, characterized above, has as shown that only a (1-fold) nonzero eigenvalue $\lambda = n$; it follows that the eigenvalue $\lambda = n$ of A is associated with the only eigenvector, having positive coordinates and determined exactly to a constant coefficient[16]:

$c \neq 0$

---

16 Ibid, p. 89.

$$w = w_1, w_2, \ldots, w_n$$

satisfying equation

$$Aw = \lambda w$$

where $\lambda = n$.

The vector w can be treated as a scale for measuring the values of objects x_i, where $i = 1, 2, \ldots, n$, being respectively rows and columns of the matrix A, for A can be interpreted as a table of the following form

	x_1	x_2	...	x_n
x_1	a_{11}	a_{12}	...	a_{1n}
x_2	a_{21}	a_{22}	...	a_{2n}
.	.	.	.	.
.	.	.	.	.
.	.	.	.	.
x_n	a_{n1}	a_{n2}	...	a_{nn}

The vector will constitute a more convenient quotient scale if we normalize its coordinates

$$w'_1 = \frac{w_1}{\sum_{i=1}^{n} w_i}$$

$$w'_2 = \frac{w_2}{\sum_{i=1}^{n} w_i}$$

$$\vdots$$

$$w'_n = \frac{w_n}{\sum_{i=1}^{n} w_i}$$

Then,

$$\sum_{i=1}^{n} w_i' = 1$$

The theoretical model sketched above shall be used to solve the problem of determining social preferences on the basis of the simple rule of majority for the pairwise comparison of alternatives, under any research condition when $n \geq 2$ (and with any number of subjects).

To this end let us consider to that end the empirical matrix A^*, whose elements are the ratios of votes in favour of each of the alternatives, in compared pairs of alternatives (alternatives in rows versus alternatives in columns), and whose properties are as it shall turn out similar to those of A discussed above[17].

$$A^* = \begin{bmatrix} g_{11} & g_{12} & \cdots & g_{1n} \\ g_{21} & g_{22} & \cdots & g_{2n} \\ \cdot & \cdot & \cdot & \cdot \\ \cdot & \cdot & \cdot & \cdot \\ \cdot & \cdot & \cdot & \cdot \\ g_{n1} & g_{n2} & \cdots & g_{nn} \end{bmatrix}$$

The elements g_{ij} stand for the ratio of the number of subjects choosing the alternative x_i compared to the alternative x_j (let us denote this by r_i) to the number of subjects choosing the alternative x_i (denoted by r_j). That is

1′ $$g_{ij} = \frac{r_i}{r_j}$$

Another property

2′ $$g_{ij} > 0$$

[17] In (Saaty and Bennet 1977, p. 238) data regarding evaluations ascribed to alternatives according to criteria of insufficient clarity. In Mirkin, in turn, although the data is countable, their definition differs from that found in the present paper (cf. footnote 12).

is an assumption (practically without any significance) regarding the distribution of preferential judgements. The fact that every element of A^* is greater than zero involves the exclusion of cases of unanimity in empirical investigations. In a given situation it can cause zero in the numeral, which makes the whole fraction equal zero. Such an elimination can be carried out in various ways. The least harmful way is probably replacing the zero resulting from the fraction $0/n$ (where n stands for the number of voters) by the fraction $1/1{-}n$. If n is large (such situations are most likely to occur in practice) then this procedure is legitimate. By doing so, we also exclude the possibility of zero appearing in the denominator of the fraction g_{ij} (located below the main diagonal of the empirical matrix A^*).

Another problem that arises in connection with the interpretation of elements of A^* (i.e. g_{ij} treated as a ratio of votes in favour of the alternative i versus the alternative j) is a lack of voters' preference with respect to a given pair of alternatives. Information generated by such subjects is not taken into account. The final result is determined only by those who express strong preferences. This is not, however, a peculiarity of this particular model because, generally, in any empirical research (concerning social life) it is only those who express their opinion that shapes the outcome, rather than those who abstain.

3′ $$g_{ij} = \frac{1}{g_{ji}}$$

4′ $$g_{ii} = 1$$

5′ $$Tr(A^*) = \sum_{i=1}^{n} g_{ij} = n$$

Property six however, requires, some comment. Elements of the empirical matrix stand for, as we have already mentioned, ratios of votes in favour of the alternative in rows versus the alternative in columns. Potential acceptance of the consistency condition would entail some restrictions on the distribution of votes in favour of respective alternatives and such restrictions would be fulfilled in practice only by exception. Therefore, in the case of the empirical matrix A^*, interpreted as above, we ought to accept the failure to fulfil the consistency condition, i.e. accept the following equation:

6′ $$g_{ij} g_{jk} = g_{ik}$$

The failure to fulfil the consistency condition entails a change of the rank of the empirical matrix A^*. By definition we know that the rank of a matrix denotes the maximum number of linearly independent columns (or rows) of that matrix[18]. But if the square matrix ($n \times n$) of the type A^* does not fulfil the consistency condition, then all n rows (columns) are linearly independent. So

7′ $$rz\, A^* = n$$

The fact that property 7′ of A^* is different from property 7° (of A) influences the magnitudes of the eigenvalues. We have to note, however, that it is still legitimate to rely on Perron's theorem and on theorem five, concerning eigenvalues of square matrices. Just like the theoretical matrix A, the empirical matrix A^* also has n eigenvalues. In the case of the former, there is only one non-zero eigenvalue (equal to the trace of the matrix, i.e. to n) while all other eigenvalues equal zero. In the case of a matrix that does not fulfil the consistency condition, there are λ_{max} magnitudes more or less close to n and there are $n-1$ eigenvalues ($\lambda_1, \lambda_2, ..., \lambda_{n-1}$) of magnitudes usually close to zero[19]. It is possible to prove that in matrices, whose elements are located under the main diagonal, are inverse to elements above it, the maximum eigenvalue λ_{max} is greater or equal to the trace of the matrix[20], i.e.

$$\lambda_{max} \geq Tr\,(A^*) = n$$

Just like in previous cases, the scale we are interested in finding (consists in the problem of the eigenvalue and the eigenvector, and it ought to satisfy the following equation:

$$A^* w^* = \lambda_{max} w^*$$

where the eigenvector w^* is calculated for the maximum eigenvalue.

---

18 (Kowalczyk 1976, p. 47).

19 (Franklin 1968). Cf. (Saaty and Bennett 1977, p. 239).

20 Ibid.

The eigenvector w^* satisfying the above equation is, as mentioned above, determined by the constant factor c and its interpretation is analogous to the case of A. The magnitudes of its coordinates will add up to 1 when normalized:

$$1 = \frac{w_1}{\sum_{i=1}^{n} w_i}$$

$$2 = \frac{w_2}{\sum_{i=1}^{n} w_i}$$

$$\vdots$$

$$n = \frac{w_n}{\sum_{i=1}^{n} w_i}$$

The resulting vector (interpreted as the scale)

$$\omega = [\omega_1, \omega_2, \ldots, \omega_n]$$

gives us results that measure the proportions of votes in favour of each of the alternatives (the higher the coordinate, the more preferred the alternative is).

At this point, there can arise the question as to whether the results obtained in the above way agree with the formula of the simple rule of majority based on pairwise comparison of alternatives. To put it in other words: do the operations involved in computing eigenvalues and eigenvectors distort the ideas implicit in the simple rule of majority? Let us consider an uncomplicated example of 30 subjects choosing between two alternatives: x_1 and x_2. Let us assume that, as a result of research, we find out that 50% of votes are in favour of x_1 and 50% are in favour of x_2 (the assumption is that everybody expressed his or her preference). This means that half of the total number of votes (voters) favoured x_1 and the other half favoured x_2. That is, neither of the alternatives is socially preferred over the other. Let us examine what results we get from applying eigenvalues and eigenvectors. Results of the research can be expressed in the form of the empirical matrix A^*.

$$A^* = \begin{bmatrix} 1 & \frac{15}{15} \\ 1 & \frac{15}{15} \end{bmatrix} \begin{bmatrix} 1 & 1 \\ 1 & 1 \end{bmatrix}$$

It turns out that the eigenvalue with the greatest absolute value equals λ_{max} = 2,000. The respective eigenvector has the form of:

$$\omega = [0{,}5000,\ 0{,}5000]$$

We notice that no change took place, for the normalized eigenvector produces a result similarly to the method previously employed social indifference with respect to both alternatives (i.e. none of them is preferred more than the other). We shall consider now a different distribution of individual preferences (choices). Let us assume that the alternative x_1 was favoured by 5 people, while the alternative x_2 by 25 people. This situation means that not only is x_2 preferred over x_1 but also that x_2 was rated five times more highly than x_1. Let us examine what the expression of this fact, in terms of coordinates of the eigenvector, is.

The empirical matrix A^* has in this case the form of

$$A^* = \begin{bmatrix} 1 & \frac{5}{25} \\ \frac{25}{5} & 1 \end{bmatrix} = \begin{bmatrix} 1 & 0{,}200 \\ 5{,}000 & 1 \end{bmatrix}$$

λ_{max} = 2,000

$$\omega = [0{,}1667,\ 0{,}8333]$$

So, the alternative x_2 is also preferred over the alternative x_1. The level of preference of x_2 versus x_1 is also maintained, for

$$0{,}1667 \cdot 5 = 0{,}8333$$

This convergence of results is impressive and in some sense optimistic because the procedure based on computing eigenvalues and eigenvectors is especially useful for investigating multi-alternative ($n > 2$) preferences. This result encourages us to extend the above model based on the matrix of comparisons A^* to multi-criterial choice, which can be promising in empirical investigation of the choice of scientific theories. Such an extension is made possible by using the notion of the hierarchical set introduced by Saaty and Bennett[21].

Definition of the hierarchical set

Any non-empty set of elements has, according to Saaty and Bennett, a hierarchical character if it can be represented in the form of mutually disjoint ordered subsets Z^1, Z^2, ..., Z^n, also called strata[22]. Each stratum occupies a determined k^{th} place in the hierarchy H (k = 1, 2, n). Each element z^i_j of each of the strata z^i is dominated by at least one element of the stratum directly above it, unless it occupies the highest position in the hierarchy H, and, conversely, each element z^i_j of each of the strata z^i dominates at least one element of the stratum directly below it, unless it occupies the lowest position in the hierarchy H. The notion of a given element of a higher stratum dominating elements of a lower stratum shall become clear when we assume that:

a) each element of each stratum is assigned some numerical value;

b) the sum of numerical values of the dominant element equals the numerical value of the dominating element of the higher stratum.

If, for example, the numerical value of the element z_1^2 from the stratum below is designated by $w(z_2^1)$ and, by analogy, the values of, elements dominated by it are designated as follows:

$$w(z_1^3), w(z_2^3), \ldots, w(z_{m3}^3),$$

then we have the following equality[23].

$$w(z_1^2) = w(z_1^3) + w(z_2^3) + \ldots + w(z_{m3}^3),$$

The authors call the hierarchy H a complete hierarchy if each element of a given stratum is dominated by each element of the stratum directly above it[24]. Then, in an incomplete hierarchy, an element of the kth stratum can be dominated only by some subset of elements of k–1 and, conversely, an element of the stratum k–1 can dominate only some subset of elements in each stratum.

---

21 Ibid., s. 240.

22 Ibid.

23 Ibid.

24 Ibid.

It is assumed for theoretical reasons that the topmost stratum consists of the set of only one element (z^l), which is assigned the value of one. To generalize the above considerations, let us put it more clearly:

$$w(z_i^k) \; by \; w_i^k$$

So for stratum two we have

$$\sum_{i=1}^{m_2} w_i^2 = w^1 = 1$$

By analogy, we can list equalities for the values of elements found on lower strata. However, we have to remember to carry out respective relativizations

$$\sum_{i=1}^{m_3} w_i^3 = w_1^2$$

$$\sum_{i=1}^{m_3} w_i^3 = w_2^2$$

$$\vdots$$

$$\sum_{i=1}^{m_n} w_i^n = w_{m_{n-1}}^{n-1}$$

The hierarchy under consideration (H) can be presented graphically as follows:

$$\begin{array}{cccc} & z^1 & & \\ z_1^2 & z_2^2 & \cdots & z_{m_2}^2 \\ z_1^3 & z_2^3 & \cdots & z_{m_3}^2 \\ & \vdots & & \\ z_1^k & z_2^k & \cdots & z_{m_k}^k \\ & \vdots & & \\ z_1^{n-1} & z_2^{n-1} & \cdots & z_{m_{n-1}}^{n-1} \\ z_1^n & z_2^n & \cdots & z_{m_n}^n \end{array}$$

Solid lines stand for the relations of the domination of an element from a higher stratum over an element from the lower stratum. Interpretation of the numerical values associated with each of the elements of strata of the hierarchy *H* depends on specific applications. In the case of social choice discussed in this present paper, they stand for the coordinates of eigenvectors, as it shall later turn out.

Let us remind ourselves of the way of constructing the empirical matrix A^* (discussed in the previous section). The matrix was created as a result of a pairwise comparison of a elements of the set. It is possible to imagine a similar procedure applied to the elements of strata (subsets Z_i of the set *Z*). After calculating eigenvectors from them, we obtain the following formula[25]:

1. $$w_j^{k+1} = c_{j1}v_1^k + \cdots c_{jm_k}v_{m_k}^k$$

where $j = 1, 2, ..., m_{k+1}$

$$v_1^k, \cdots, v_{m_k}^k$$

are coefficients of the eigenvector calculated from the matrix of comparisons B^* for the stratum

$$z_1^k, z_2^k, \cdots, z_{m_k}^k$$

$$c_{j1}, \cdots, c_{jm_k}$$

are elements of jth row of the matrix *C*, whose subsequent columns from 1 to m_k constitute eigenvectors of matrices

$$D_1^*, D_2^*, \cdots, D_{m_k}^*$$

obtained for elements of the stratum $k + l$, i.e. for

$$z_1^{k+1}, z_2^{k-1}, \cdots, z_{m_{k+1}}^{k+1}$$

by means of comparisons with respect to the subsequent elements from the stratum *k*.

---

[25] Ibid, p. 241.

That is a comparison (as in the above empirical matrix A^*) of the elements with respect to the element $z^1{}_k$ (being here the criterion of comparisons) and results in the matrix $D_1{}^*$. Comparison of the elements with respect to the element $z^2{}_k$ (being here the criterion of comparisons) and results in the matrix $D_2{}^*$. Comparison of the elements with respect to the element z^k_{mk}, which results in the matrix $D_r{}^*$ and so on, until we have compared all the elements.

Saaty and Bennett generalize formula one for the complete hierarchy *H*, and it takes the following form[26]:

2. $$w_{i_k} = c_{i_k} v_{1_{k-1}} + c_{i_k 2} v_{2_{k-1}} + \cdots + c_{i_k m} v_{m_{k-1}}$$

Hierarchical models of social choice

Utilizing the conceptual apparatus of the theory of social choice enables the formulation of three models of social choice, namely: one-dimensional choice, multi-dimensional choice with fixed criteria individual preferences, and of multi-dimensional choice with varying criteria of individual preferences. It is worth noticing that a procedure similar to the model of one-dimensional choice is fundamental to the case analyzed by B. G. Mirkin and that the model of multi-dimensional choice with fixed criteria of individual preferences is fundamental to Saaty and Bennett's analyses. These authors, however, analyse data of a type different from that of the present study, which completely changes not only the empirical meaning of the outcome but also the structure of their models. The following reconstructions of each of the models are, as we have already mentioned, elaborations of the simple rule of majority based on the pairwise comparison of alternatives and they are its modifications to fit three possible cognitive purposes of empirical research on the choice of scientific theories.

The model of one-dimensional choice

The model of one-dimensional choice, for the situation of $N = \{i_1, i_2, ..., i_n\}$ voters choosing from among $S = \{x_1, x_2, ..., x_m\}$ can be presented as the hierarchy of three strata:

---

[26] Ibid.

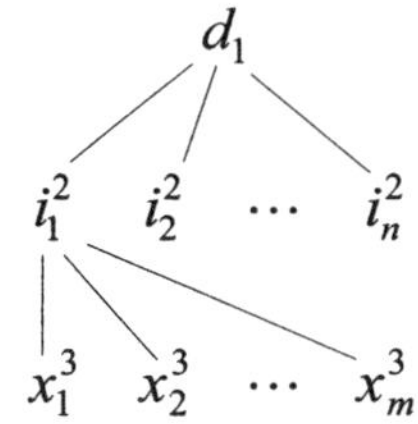

The third stratum contains the set of alternatives $S = \{x_1, x_2,..., x_m\}$. The second stratum contains the set of voters $I = \{i_1, i_2,..., i_n\}$ and the first stratum, consisting of one element and being practically without significance, can be interpreted as the totality of chances in voting of particular voters (i.e. members of the community under investigation) whose preferences with respect to the alternatives $x_1, x_2,..., x_n$ we are interested in.

In empirical research, as well as in democratic systems, individual preferences of all members of a given group are treated equally, as a rule. The model under consideration expresses this by means of ascribing each of the voters (members of the community making a choice) $1/n^{\text{th}}$ of influence on the final outcome of the ballot. That is, stratum two is represented by the vector:

$$u = [u_1, u_2, \cdots, u_n]$$

where the coefficients are

$$u_1 = u_2 = \cdots = u_n = \frac{1}{n}$$

The final result, i.e. the normalized eigenvector for the 3rd stratum, that is for the set of alternatives $S = \{x_1, x_2,..., x_m\}$ is obtained in the way described in the first part of the present section. Individuals make their choice by comparing pairs of alternatives, which constitutes the empirical matrix A^*.

$$A^* = \begin{bmatrix} g_{11} & g_{12} & \cdots & g_{1m} \\ g_{21} & g_{22} & \cdots & g_{2m} \\ \cdot & \cdot & \cdot & \cdot \\ \cdot & \cdot & \cdot & \cdot \\ \cdot & \cdot & \cdot & \cdot \\ g_{m1} & g_{m2} & \cdots & g_{mm} \end{bmatrix}$$

The eigenvector calculated for the greatest eigenvalue λ_{max} of A^* takes the following form:

$$w^* = w_1^*, w_2^*, \cdots, w_m^*$$

After normalizing w^* we shall obtain the scale:

$$\omega = [\omega_1, \omega_2, \cdots, \omega_m]$$

which constitutes, at the same time, the desired social ordering of alternatives

$$S = \{x_1, x_2, \cdots, x_m\}$$

The model of one-dimensional choice is based on the simplest hierarchy of the three strata, while the 1st stratum is, for the reasons mentioned above, disregarded in practice. Hence, only the two remaining strata are meaningful. The 2nd stratum is represented by the vector of n identical coordinates of magnitude $1/n$ and we are ultimately interested only in the ordering of alternatives contained in the normalized vector, and therefore this stratum is also disregarded in mathematical calculations. Therefore, A^* formulated as a result of the pairwise comparison of the alternatives x_1, x_2,..., x_m by voters i_1, i_2,..., i_m is the only matrix occurring in this model and the eigenvector computed on its basis is the final result of our investigation.

The model of one-dimensional choice is relatively simple while models multidimensional choice are more complicated.

The model of multidimensional social choice with fixed criteria of individual preferences

The hierarchy H for a model of multidimensional choice with fixed criteria of individual preferences consists of four strata:

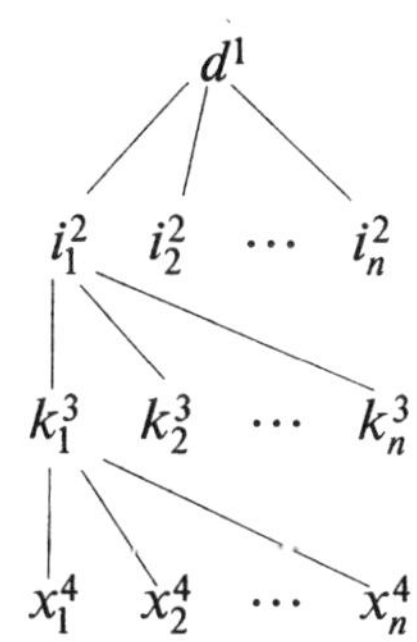

To sets of voters and of alternatives, we add the set $K = \{k_1, k_2, \ldots, k_r\}$ of criteria of choice. It is placed after the set of votes and it constitutes the 3rd stratum, which is located above the set of alternatives (i.e. the 4th stratum). These criteria are fixed for all voters and they are a basis of evaluation of alternatives by voters. There can be any number of them.

There are two stages in determining the final outcome consisting in the social ordering of alternatives. First, the subjects arrange the criteria of choice in order of decreasing importance (for a problem, a given branch of science, or for a given type of scientific theories, etc.). This process results in one empirical matrix B^*.

$$B^* = \begin{bmatrix} b_{11} & b_{12} & \cdots & b_{1r} \\ b_{21} & b_{22} & \cdots & b_{2r} \\ & \cdots\cdots\cdots & & \\ & \cdots\cdots\cdots & & \\ & \cdots\cdots\cdots & & \\ b_{r1} & b_{r2} & \cdots & b_{rr} \end{bmatrix}$$

The eigenvector computed on its basis shall be represented by

$$v=\left[v_1,v_2,\cdots,v_r\right]$$

The next stage consists in arranging alternatives separately, with respect to each of the criteria. As a result, we obtain as many empirical matrices as there are criteria of choice: A_1^*, A_2^*,..., A_r^*. Eigenvectors are computed from each of the empirical matrices, i.e. from B^* and from A_1^*, A_2^*,..., A_r^*. The eigenvector for the set of alternatives (the 4th stratum) is obtained according to the formula:

$$w_j^*=\left[c_{j1}v_1+c_{j2}v_2+\cdots c_{jr}v_r\right]$$

where j = 1, 2,..., m; v_1, v_2,..., v_r are coefficients of the eigenvector v, calculated from the empirical matrix B^*, for the stratum containing the set of criteria $K = \{k_1, k_2, ..., k_r\}$; c_{j1}, c_{j2}, ..., c_{jr} are elements of the jth row of C^*, whose columns, numbered from 1 to r, are eiganvectors of empirical matrices A_1^*, A_2^*,..., A_r^*. After normalizing the eigenvector

$$w^*=\left[w_1^*,w_2^*,\cdots,w_m^*\right]$$

obtained in the above way we obtain the desired vector:

$$\omega=\left[\omega_1,\omega_1,\cdots,\omega_m\right]$$

from which we obtain the social ordering of alternatives.

The model of multidimensional social choice with varying criteria of individual preferences

This model takes into consideration not only the fact of multiplicity of criteria of individual orderings but also of a diversity of the subjects' criteria of preferences. The hierarchy H for this model consists of four strata, just like in the previous case. However, not all the criteria k_1, k_2,..., k_r are bases of choice for every voter, for it often happens that people differ with respect to the relevance of different criteria of evaluations. Some criteria are considered very important, while others are thought to be quite insignificant. On the one hand, the insignificant criteria ought not, as a rule, be taken into consideration in the ordering of alternatives. On the other hand, there is no agreement as to what criteria are important and what are insignificant. There are controversies, e.g. about ideological criteria and criteria connected with people's outlooks on life and, in the case of choice of scientific theories, also of axiological and methodological ones.

In relation to the above, it seems justified that the criteria people follow while determining individual preferences should be arbitrary. Therefore, using this model to study n voters, one singles out n non-disjoint subsets of the set of criteria.

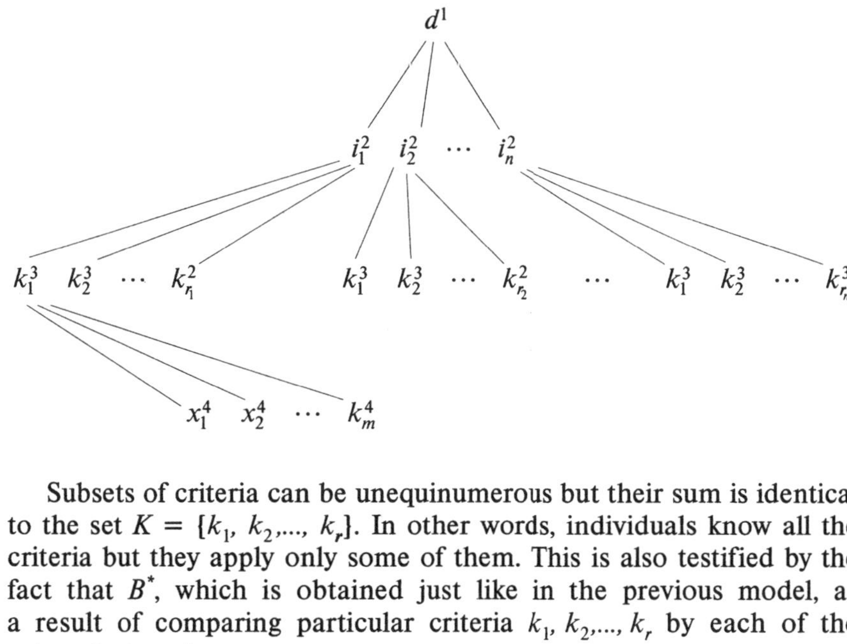

Subsets of criteria can be unequinumerous but their sum is identical to the set $K = \{k_1, k_2, \dots, k_r\}$. In other words, individuals know all the criteria but they apply only some of them. This is also testified by the fact that B^*, which is obtained just like in the previous model, as a result of comparing particular criteria $k_1, k_2, \dots, k_r$ by each of the voters $i_1, i_2, \dots, i_n$.

From the fact of freedom in choosing the criteria it follows that some can be bases for individual preferences more often than others. In particular, it may happen that everybody chooses only one criterion, which will lead to a convergence of this model with the model of one-dimensional choice. It is also possible that some criteria (criterion) will be unpopular, i.e. that nobody, or a very small number of voters uses them to establish their own orderings. We can then safely assume that such criteria (criterion) will also be valued law in the social ordering calculated from B^*. As a result, we can eliminate from further calculations the criterion (criteria) of clearly low popularity. We have to bear in mind, however, the danger of distorting the final outcome, i.e. the social ordering of alternatives.

As a result of pairwise comparisons of alternatives $x_1, x_2, \dots, x_m$ with respect to the subsequent criteria, we obtain the matrices $A_1^*, A_2^*, \dots, A_s^*$. The number of these matrices can differ from r (which stands for the number of criteria) if we have eliminated some of them, i.e. when $s \leq r$. We then proceed as before, i.e. we compute the eigenvector for the set of alternatives according to the formula:

$$w_j = c_{j1}v_1 + c_{j2}v_2 + \cdots + c_{js}v_s$$

where $j = 1, 2, \dots, m$; $v_1, v_2, \dots, v_s$ are coefficients of the eigenvector w, calculated from the empirical matrix B^* for the stratum containing the set of criteria $K = \{k_1, k_2, \dots k_r\}$ (an elimination of some criteria is possible only later, while processing the results of an investigation, but before computing the value of w_j, and that is why B^* comprises r criteria, i.e. all criteria that have been taken into consideration in particular research but not necessarily merely devised by the researcher, for the set of criteria $k_1, k_2, \dots, k_r$ can be determined after preliminary survey of the subjects): $c_{j1}, c_{j2}, \dots, c_{js}$ are elements of the jth row of C^*, whose subsequent columns *1* to *s* are eigenvectors of empirical matrices $A_1^*, A_2^*, \dots, A_s^*$.

Having normalized the coefficients of the eigenvector:

$$w^* = \left[w_1^*, w_2^*, \cdots, w_m^*\right]$$

which represents the social ordering of alternatives.

The procedures in each of the models are relatively simple, especially because of the application of computers in statistical calculations. Let us summarize the main steps of the research process.

In the case of the one-dimensional model there are the following stages:

1) comparing alternatives: the results are recorded in the form of the empirical matrics A^*;
2) computing the eigenvector w^* from the matrix A^*;
3) normalizing the eigenvector w^*.

In the case of the multi-dimensional model with fixed criteria the procedure consists of six steps:
1) comparing criteria: the result is recorded in the form of the matrix B^*;
2) comparing alternatives, with respect to all of the accepted criteria (the result for each criterion is recorded as a matrix, leading to k empirical matrices $A_1^*, A_2^*, \ldots, A_k^*$);
3) computing the eigenvector v from the matrix B^*;
4) computing eigenvectors $c_1, c_2, \ldots, c_k$, whose coordinates are elements of the matrix C^* from matrices $A_1^*, A_2^*, \ldots, A_k^*$;
5) computing the eigenvector w^* for the set of alternatives;
6) normalization of the eigenvector w^*.

And finally, the procedure for the model with varying criteria of individual preferences is the following:
1) comparing criteria: the result is recorded in the form of the matrix B^*;
2) comparing alternatives; with respect to the criteria chosen by the voter, the result is recorded as s for empirical matrices $A_1^*, A_2^*, \ldots, A_s^*$; if no elimination of criteria took place, then $s = k$;
3) computing the eigenvector v from the matrix B^*;
4) computing eigenvectors $c_1, c_2, \ldots, c_s$, whose coordinates are elements of the matrix C from matrices $A_1^*, A_2^*, \ldots, A_s^*$;
5) computing the eigenvector w^* for the set of alternatives;
6) normalization of the eigenvector w^*.

A common feature of these models is the fact that they rest on individual preferences and that the process of constructing social orderings goes beyond the simple method of majority. Of indubitable merit to our conceptual apparatus to conceptions of social choice that comprise of three possible situations in voting: one-dimensional choice, multidimensional choice with fixed criteria, and multi-dimensional choice with varied criteria of individual preferences. Thanks to the fact that all mathematical operations are carried out only on the numbers of results of pairwise comparisons of alternatives (although the problem of intensity of preference is not taken into account) the final outcome is obtained in the quotient scale, from which we obtain the social ordering of alternatives. We are then guided by the magnitude of a given coefficient: the higher it is, the more preferred the corresponding alternative is (this alternative bears the same number as a given coefficient).

Department of Philosophy
Maria Curie-Skłodowska University
M. Curie-Skłodowskiej 4
20-031 Lublin
Poland

REFERENCES

Amsterdamski, S. (1983). *Między historią a metodą.* Warszawa: PIW.
Arrow, K.J. (1951). *Social Choice and Individual Values.* New York: Wiley.
Borda, J.-Ch. (1781). *Memoire sur les Elections au Scrutin.* English translation: Grazia, A. (1953).
Bordes, G. (1975). Consistency, Rationality and Collective Choice. *Review of Economic Studies* **43**, 451-457.
Condorcet, J.A. (1785). *Essai sur l'Application de l'Analyse à la Probabilité des Décisions Rendues à la Pluralité des Voix.* Paris: NE.
Franklin, J.N. (1968). *Matrix Theory.* New York: Englewood Cliffs.
Grazia, A. (1953). Mathematical Derivation of an Election System. *Isis* **44**, 42-51.
Kowalczyk, B. (1976), *Macierze i ich zastosowania* [Matrixes and their Applications]. Warszawa: WNT.
Mirkin, B.G. (1974). *Problema grupowogo wybora* [The Problem of Collective Choice]. Moskwa: Nauka.
Ogryzko-Wiewiórowski, H. (1986). *Problem racjonalności wyboru społecznego.* [The Problem of Rationality of Social Choice]. Lublin: Wydawnictwo UMCS.
Opial, Z. (1976). *Algebra wyższa* [The Higher Algebra]. Warszawa: PWN.
Richelson, J. (1978). Some Further Results on Consistency, Rationality and Collective Choice. *Review of Economic Studies* **45**, 342-353.
Saaty, T.L. and Bennett, J.P. (1977). A Theory of Analytical Hierarchies Applied to Political Candidacy. *Behavioral Science* **22**, 237-245.

Poznań Studies in the Philosophy of the Sciences and the Humanities
2000, Vol. 72, pp. 133-151

Kazimierz Jodkowski

IS THE CAUSAL THEORY OF REFERENCE A REMEDY FOR ONTOLOGICAL INCOMMENSURABILITY?

ABSTRACT. The paper analyses a criticism of ontological incommensurability set forth by proponents of the so-called 'causal theory of reference'. According to this theory, reference of certain terms is independent both of the knowledge of those who use these terms, and of their meanings. Some thinkers hope that once the causal theory of reference is extended to deal with theoretical terms, it is going to guarantee ontological commensurability. The author argues, however, that this hope is illusive and bases his argument on analyses of concepts like "water" in the 17th century chemistry and "planet" in Ptolemaic astronomy. These examples show that the role of theory is unavoidable in the process of linking terms with reality, which makes it possible for ontological incommensurability to appear.

0. Introduction. Ontological incommensurability according to P.K. Feyerabend and T.S. Kuhn

Philosophers of science express extreme opinions on incommensurability. A great deal has been written about the thesis of incommensurability of successive theories. Many philosophers point out the complex and multidimensional character of the thesis as expressed in the writings of its chief advocates: Thomas S. Kuhn and Paul K. Feyerabend.

Scientists accepting different theories can perceive the world in different ways. If the theories they accept are incommensurable, then their observations will also be radically different. A change of beliefs can be accompanied by a sudden, complete change of observations, analogous to a *Gestalt* switch. The linguistic dimension of the incommensurability thesis is the best known. In the history of science we can find cases of changes of theories such that both theories, the earlier and the later, cannot be translated into each other in an exact sense. The methodological dimension of the incommensurability thesis boils down to the claim that in the process of switching from one theory to another, the methodological "superstructure," i.e. the criteria distinguishing science from non-science as well as the standards of evaluation, can also change. Science can, at least to some degree, take up different problems and evaluate their solutions in different ways[1].

There is also the ontological dimension of the incommensurability thesis.

A scientific theory in Feyerabend's sense comprises of an ontology that determines what exists (cf. Feyerabend 1975, pp. 176-177).

> . . . even everyday languages, like languages of highly theoretical systems, have been introduced in order to give expression to some theory or point of view, and they therefore contain a well-developed and sometimes very abstract ontology (Feyerabend 1962, p. 76).

Therefore, a change of a universal theory must be accompanied by a change of ontology. Feyerabend even maintains that a complete replacement of the ontology must take place.

> What does happen is, rather a *complete replacement* of the ontology (and perhaps even the formalism) of T' by the (ontology and the formalism) of T and a corresponding change of the meanings of the descriptive elements of the formalism of T' (provided these elements and this formalism are still used) (Feyerabend 1962, p. 29; see also Feyerabend 1962, p. 59).

Obviously, such a strong statement of Feyerabend's, talking about a complete replacement of the ontology instead of a change of it, requires some justification (see Jodkowski 1984, chapter II, esp. pp. 47-50).

What is the situation described by Feyerabend like? Every alternative scientific theory contains its own ontology. These ontologies are different; therefore, as a theory changes the earlier world in some sense disappears: ". . . the earlier visual world . . . literally *disappears*" (Feyerabend 1975, p. 228).

So it turns out that objects which previous theories discussed never existed.

---

[1] I characterized the incommensurability thesis presented in the writings of Kuhn and Feyerabend in my two books: (Jodkowski 1984, pp. 8-34) and (Jodkowski 1990, pp. 328-379).

> Scientific laws can be revised, they often turn out to be not just locally incorrect but entirely false, making assertions about entities that never existed (Feyerabend 1975, p. 188).

> . . . we very often discover that entities we thought existed did, in fact, not exist (Feyerabend 1965a, p. 170).

As a result concepts of rejected theories would have no extension (cf. Feyerabend 1962, p. 74; and Feyerabend 1965b, p. 268).

Kuhn pointed also to ontological variability. One of components of the paradigm in wider sense, or the so-called disciplinary matrix, are ontological models dealing with the structure of the world (see Jodkowski 1981, p. 59 and Jodkowski 1990, Part II, chapters 1 and 3). During paradigm change

> . . . it is not only the forms of the laws that have changed. Simultaneously we have had to alter the fundamental structural elements of which the universe to which they apply is composed (Kuhn 1970a, p. 102).

> Successive paradigms tell us different things about the population of the universe and about that population's behavior. They differ, that is, about such questions as the existence of subatomic particles, the materiality of light, and the conservation of heat and energy. These are the substantative differences between successive paradigms, and they require no further illustration (Kuhn 1970a, p. 103).

Before the theory of relativity the term "mass" did not refer to the same physical quantity as it does today (cf. Kuhn 1970a, pp. 101-102)[2].

We can distinguish in Kuhn's works two kinds of ontological variability. As they are not introduced and analyzed explicitly, the easiest way to identify them is to refer to Kuhnian distinction between translation and interpretation.

In his earlier works, apart from referring to methodological values, Kuhn mentioned an imperfect way of making positions closer to each other and discussing difficult issues of choice between theories. Incommensurability of scientific terms does not have to lead to a deadlock. It can be avoided by the mutual translation of alternative theories into the language of a given theory. In order to accomplish this, it is necessary to gain a profound knowledge of the essence of the language of science, investigate how it relates to the world, how scientists gain knowledge of these relations and how they are conveyed to successive generations of a given scientific community. A breakdown of communication occurring during a revolution is never total nor beyond discussion. Most of the terms common to both theories function in them

---

[2] A similar argument can be found in Quine, who considers the case of the neutrino (Quine 1960, p. 16).

in the same way. Problems with translation arise only for some of the terms and statements containing them. The rest of the terms, which keep their meanings during the switch to a successive theory, provide a sufficient basis to analyse the differences between the theories. Incommensurability does not mean total incomparability and untranslatability (cf. Kuhn 1970b, p. 202; Kuhn 1970c, p. 267).

A scientific group centred around a paradigm is a language community. All communities share an everyday language and most of the language of science. Translation of a theory by members of a different language community into their own language is sufficient, according to earlier works of Kuhn, to understand the theory. However, this is neither a complete translation nor a complete understanding. It is undetermined in Quine's sense. In slicing the world in different ways, languages must assume some theory. This is the source of problems in translating different theories. We learn a language when we get to know the world that it describes. The same process takes place during scientific training, whilst getting to know paradigmatic solutions to problems. The job of the language of science is not to describe just any world but the world as it is (or, rather, as it thought is to be). It is difficult to express in this language what is not expected to occur and this is precisely the situation that exists when comparing different theories. They predict that which is impossible according to another theory. Here a strict connection of the language and ontological dimensions of incommensurability is revealed.

Difficulties in communication regard, then, not only language but also nature itself (or, rather, a picture of nature). Therefore translation is only an imperfect means of establishing contact between groups. Understanding is only partial. Full understanding is only possible once a new paradigm has been accepted.

The idea of partial translation is merely sketched in Kuhn's earlier works and is tacitly abandoned in his recent writings. Connections between the Quinean thesis of indetermination of translation with the conception of incommensurable paradigms have also been merely mentioned. It is not certain whether these connections are correct. Quine's thesis seems to assume, for example, a common basic ontology for all languages. It is *the same* world being sliced in different ways. Different paradigms, however, occasionally seem to postulate the existence of different objects, besides common entities.

In his more recent writings Kuhn thinks that the category of meaning might not be the right or the best category to analyse the problems of incommensurability. To understand earlier theories the procedure of interpretation, in addition to translation, is of essential importance. In earlier works he did not write about it because he had accepted a mistaken, as he thinks now, supposition of Quine's that interpretation is some kind of translation (in Quine's terms "radical translation"). Many of Kuhn's critics showed that the procedure of interpretation (which they also equate with translation) is successful, and this in their view speaks against any version of incommensurability (e.g. Davidson 1973-1974, p. 19; Kitcher 1978, pp. 519-529; Putnam 1981, p. 116). Kuhn, however, maintains that it is necessary to carefully distinguish the two processes to avoid confusion.

Translation is carried out by somebody who knows both languages. Having encountered a text written in one of them, he replaces words or strings of words by appropriate expressions of the other language with the intention of finding an equivalent text. However, linguists argue as to what the "equivalent text" actually is. Usually, they mention a realization of the postulate of meaning and referential identity.

One has to pay special attention to two features of translation. Firstly, the process of translation does not itself cause changes in meanings of the target language. This feature results from the fact that both languages exist before translation of a text is undertaken. Secondly, a translation does not (or, rather, should not) contain any additional expressions due to the translator, e.g. comments, introductions, etc. The outcome of an ideal translation should be a simple replacement of words of one language by those of another.

The researcher performing interpretation (e.g. a historian or an anthropologist) does not have to know both languages right at the very beginning. So, the text appears to him as a system of inscriptions or sounds without meaning. Then, the interpreter looks for their sense by observing the behaviour of senders and receivers of the text, and the circumstances of expressing it. He creates hypotheses attributing sense to the text under given circumstances. A successful interpretation is different from a correct translation of the text. It is a part of learning a new language, getting to know it, or mastering it.

A correct interpretation of some utterance in one language does not influence the possibility of translating it into the interpreter's own language. If s/he tries to describe in his/her native tongue the reference of the utterance s/he has successfully interpreted, and if s/he is entirely successful, then in such a case incommensurability is out of the question. This is then a complete and accurate translation.

Analysing the Quinean example of "gavagai," Kuhn writes that the interpreter can undertake an attempt to describe in his/her own language what the word refers to. If such a description fits all and only those creatures whose appearance evokes in natives the utterance of "gavagai," then the description is the desired translation and the word "gavagai" can be introduced into the interpreter's language as an abbreviation of this description.

However, in Kuhn's opinion, this situation is not, a rule as it is not always true. If in the process of interpretation one finds features of reality that have no corresponding words in the interpreter's language then complete translation is not possible. It is so, Kuhn says, if the language being interpreted structures the world in a different manner. Kuhn reserves the term "incommensurability" exclusively for such circumstances.

This opinion is worth some consideration. Why does one occasionally not find appropriate words to describe the reference of statements of a different language? Kuhn's explanations, based on getting to know new properties and a different structuring of reality, are not too clear. In the process of performing normal science one also gets to know new properties and new objects, while incommensurability is out of the question there and complete translation can always take place. When do new properties and different structuring lead to incommensurability? What does it depend on?

At this point, a weakness of Kuhn's analysis of the problem becomes evident. Although he is surely right in his identification of the situation where a translation can sometimes be found and sometimes not, it is his explanation of this state of affairs that is insufficient.

Kuhn's opinion can be stated more precisely. One can say that incommensurability is revealed if in the process of interpretation of some term there are not only no appropriate words in the interpreter's own language, as Kuhn says, but the creation of appropriate words is also not possible. Such an addition, however, inevitably leads to another question: Why can appropriate words be created on some occasions and not on others?

A different structuring of the world can be conceived in two ways. Firstly, by dividing the same world as before (a new theory has, then, the same basic ontology as the previous one)[3]. Secondly, by postulating (sometimes apart from the ones accepted so far) new objects or new properties, events, relations, etc., which in some sense could not exist under the previous way of structuring the world. In the first case it is always possible to create in the language of a new theory expres-

---

[3] Cf. (Kuhn 1987, p. 19). The text appeared earlier as the Occasional Paper #18, Center for Cognitive Science, Massachusetts Institute of Technology, Cambridge, Mass. 1981. In this paper Kuhn discusses redistribution of the same objects into different natural kinds.

sions that are translations of expressions of the old theory. In the second case the result of new structuring is a partially, or in an extreme case a totally, different world.

Which type does Kuhn mean? As I mentioned above, Kuhn excludes the existence of incommensurability if it is possible to describe in one language the references of expressions of a different language. Therefore, when he talks about different structuring of the world, he probably means only the second of our types. For the possibility of finding a complete translation would be excluded only if a new theory either postulated new kinds of objects, phenomena, events, etc. (they are new in the sense that not only do they not exist according to the ontology of the old theory but they also cannot exist according to it) or if it denied the possibility of objects previously postulated by a competing theory, i.e., if it rendered the referential claims of some of its terms invalid. Standard examples of the radical kind of change are the switch from classical physics using the concept of the ether to relativistic physics, which is not using it, and the switch from the chemistry of phlogiston to Lavoisiere's chemistry.

The ontological dimension of incommensurability would be that after a scientific revolution scientists would feel as if they had been transferred to a different planet or to a different world (this is Kuhn's expression). It would also be true that after a scientific revolution the whole of the earlier world in a sense disappears (this is Feyerabend's expression). This is a denial of the hitherto cherished idea of constant progress in the history of science and it is no wonder that philosophical circles launched a fierce critique of the idea of incommensurability of scientific theories. It was pointed out that if incommensurable theories concern different things then they cannot compete for they are simply different theories, for example Keynes' theory of economics and the quantum theory in physics. It was maintained that if theories do compete then they must, in some sense, concern the same domain. The appearance of the incommensurability thesis was judged to be the result of a change in scientific views. A sufficiently substantial change of scientific opinions was held responsible for the view that theories before and after a scientific change concern something different. Among critical voices there are voices hinging on the so-called causal theory of reference, also called the direct or new theory of reference. They argue for the maintenance of the same reference throughout scientific revolutions.

1. A critique of incommensurability based on the causal theory of reference

The ideas of the causal theory of reference have been articulated since the early 1960s. They were originally applied to proper names and later extended to other names, including theoretical names. The main authors are Hilary Putnam and Saul Kripke.

Kripke criticized the view that the name "Julius Caesar" is a synonym of the description "the Roman leader who crossed the Rubicon, defeated Pompeius...", etc. For if it were so, Julius Caesar would have to have crossed the Rubicon, have defeated Pompeius..., etc., and this is not true. Even his name could have been not Julius but, for instance, Marcus. According to Kripke, proper names (and other names as well) must relate to reality in some different way: not through descriptions but directly. But if it is so, then a change of our opinion with regards to the subject being named is irrelevant. The name "Nixon" refers to Nixon, no matter what people say and think about him.

In Putnam's opinion, every developed society adopts the so-called language labour division. Since nobody is an expert in every domain, laymen rely on experts. Everybody knows how to use words like "chair" or "spoon" but not everybody can talk freely about quarks, electrons, or black holes. But not only physicists talk about those objects. It is possible because laymen use experts' conclusions thanks to language cooperation, in which some people rely on others' competence. I can, for example, talk about gold reserves at Fort Knox although I cannot distinguish real gold from ersatz gold. But I can do this relying on the expertise of others.

The fact of language cooperation between experts and laymen proves that reference does not depend on the knowledge that people using a given term have. It does not depend on their knowledge of either theory or meaning of the word. Experts and laymen do have slightly, at least, different knowledge of, for example gold (some laymen even more than slightly) but in spite of this, all of them talk about the same. Just the fact that reference is independent of theory and of word meaning has caused some critics to regard the causal theory of reference as the foundation upon which to rest their critique of incommensurability. For the incommensurability doctrine of Kuhn and Feyerabend is popularly understood as resulting from a belief that radical conceptual changes take place in the history of science.

An example given by Putnam might help us understand how such a critique goes. Let us imagine we have just found out that all cats are robots and have always been robots, but this fact only came to light recently. What will our attitude toward this discovery be? Will we say that there are no cats on Earth and there have never been, or that, contrary to our previous beliefs, they are not animals but remotely controlled robots? Of course, the latter. We know such cases from the history of science. When it turned out, contrary to popular belief, that whales are not fish but mammals, nobody started to say that there had never been whales on Earth but, rather, that they are mammals.

Let me give another example. Two friends are talking. Suddenly, one of them carefully examines the collar of the other's shirt and asks, "Tell me, is your shirt white or grey?" "White," says he, "Because if it were grey, it'd be black."

All these stories prove that even a radical change of opinion does not influence the reference of the terms "cat," "whale," or "white shirt." This fact is a temptation to claim on the basis of the causal theory of reference that after radical changes of scientific theories references of terms of the theory maintain their previous reference, or that there is no ontological incommensurability. One has only to extend the causal theory to theoretical terms; for it was formulated using words causing no problems, such as "gold," "tiger," etc.

There are many ways to understand what theoretical terms are[4]. Traditionally, they were understood as terms that cannot be defined ostensively or terms whose designata cannot be observed directly. (The twentieth-century philosophy of science gradually reached the conclusion that all terms are more or less theoretical. I shall skip this fact for a moment and return to it while discussing the term "water," which appears to be more problematic than proponents of the causal theory of reference thought.) If we understand theoretical terms in such a traditional fashion, then their reference can be defined as anything that causes "such and such" observable results. For example, according to Kripke light is anything in the world that influences our eyes in a certain way, while heat is anything that produces in us a sensation called the feeling of warmth. Proponents of the causal theory of reference maintain that if we define, in the above way, references of terms belonging to theories which are, according to Kuhn and Feyerabend, incommensurable, then we can reject ontological incommensurability.

---

[4] Cf. (Jodkowski 1980a, 1983) and the bibliography of the subject which can be found there.

2. Are Kuhn and Feyerabend descriptivists?

Critics depict Kuhn and Feyerabend as descriptivists, i.e. as saying that reference is determined by the description of an object associated with its name. Consequently they are to be proponents of the idea that changes of meanings of scientific terms involve changes in ontology. For example R. Nola thinks that Kuhn accepts the so-called cluster theory of reference (to the effect that the reference of the term *t* is the object satisfying a cluster of descriptions) and that a term loses its reference if all, or a sufficient number of descriptions, fail to be satisfied by the object (cf. Nola 1980 p. 335).

But the truth is that neither Kuhn nor Feyerabend subscribes to such a form of descriptivism. They consider that changes in ontology that can properly be called incommensurability take place relatively rarely and they do not accompany any change of meaning. Feyerabend is in this respect even more "conservative" that Kuhn for Feyerabend does not regard as incommensurable some cases of change in science listed by Kuhn as transitions to incommensurable theories (e.g. the Copernican revolution) (cf. Feyerabend 1975, p. 114). Both of them, however, acknowledge the existence of scientific disputes regarding the same objects. Besides, for both of them there is a sense in which ontologically incommensurable theories do refer to the same thing, for otherwise it is clear they would be incapable of competing[5].

Feyerabend declares the occurrence of incommensurability only if the hitherto obtaining extensions of scientific terms turn out to be empty or if the hitherto obtaining extensions are abolished and a new system of extensions is proposed (cf. Feyerabend 1965b, p. 268).

Kuhn makes an appearance of incommensurability dependent on giving up the parts of a theory (e.g., its laws) used to introduce concepts of the theory (cf. Kuhn 1986). He subscribes to the view that concepts of a theory are not introduced through distinct definitions but during training involving simultaneous introduction of scientific

[5] Nola, having found a passage in Kuhn to the effect that Einstein's theory can only be accepted if Newton's theory is believed wrong, subsequently demonstrated his complete lack of the skill of interpretation. If we accept the cluster theory of reference, then Newton's and Einstein's theories refer to different objects of their own and they express analytically true statements concerning them. But in such a case it is not possible that one theory is right and the other is wrong (cf. Nola 1980, pp. 337-338). Nola is simply unable to comprehend that Kuhn can accept a sense in which two ontologically (or referentially in his terminology) incommensurable theories can still refer to the same thing.

laws, terminology and workings of the world. It takes place during apprehension of so-called exemplars.

3. An increasing role of theory in extending the causal theory of reference

An advantage of the causal theory of reference in its primary version concerning proper names (like "Julius Caesar") was that reference did not depend on meaning, description, or somebody's views. It was to be determined by the initiating causal link when somebody named an observed object with a given name for the first time and it was then conveyed to other people.

However, if this theory is extended to cover also generic names (like "tiger"), names of substances (like "gold"), and to theoretical terms, then a theory begins to play a role in determining reference. For if some standard examples are given a certain name, then it is assumed that the name can be used to name other examples of the same kind having *the same nature* (cf. Putnam 1983, p. 71). We might know the nature imprecisely but it is because of the same nature and not of some criteria that objects belong to the same kind. Using a name introduced earlier to cover other examples of the kind depends, however, on our present knowledge of the nature. For example, green lemon is classified as a kind of lemon, while a yellow fruit looking like a lemon might not be classified as lemon because the nature of lemon is determined not by fruit's color but by its DNA (cf. Putnam 1983, p. 73). Obviously, as our knowledge improves and changes, proponents of the causal theory of reference allow for the possibility of correcting the previously determined extension.

This role of knowledge becomes more and more important as the term is of increasingly theoretical character. Sometimes knowledge turns out to be decisive. Reference of the term "phlogiston" belonging to early chemistry can be defined in accordance with the causal theory in the following way.

Reference of the term "phlogiston" is any object causing:

a) calcination of metals in the air, and

b) departing from metals in the process.

Conditions a) and b) determine the nature of phlogiston. It is not sufficient to consider only the first condition. If we did so, we would have to admit that introducing a new chemical term, "oxygen", was a mistake and that, instead of "oxygen", chemists ought to use the term "phlogiston" because oxygen is responsible for what they call calcination of metals. We would only have to attach to phlogiston different theoretical opinions, which ought to be no problem for a proponent of the causal theory of reference. However, in explaining in this way why the concept of "phlogiston" vanished from chemical terminology (cf. Skalski 1991, pp. 60-61) we make the decisive role of theory in determining the reference of this term explicit. Point b) has a very theoretical character; it describes no observable phenomenon or process.

The case of "the ether" is similar and it differs only in that its chief function (of being the medium for waves of light) is not observable and we can only observe its secondary functions (e.g., causing the positive outcome of the Michelson-Morley experiment, which was, in reality, not obtained).

What conclusion can be drawn from these examples? They show that theory was essential in the process of establishing relations between the theoretical terms and nature (at least in the above cases). If this were a rule, rather than an exception, then claims of opponents of ontological incommensurability to have refuted the thesis of discontinuity in the development of science would be strongly contested. For if reference depends also on theory is it, then, preserved once theory is completely abandoned?

4. Examples of ontological incommensurability and the causal theory of reference

I would like to return to the issue which I touched upon earlier, namely that all terms of science (and of everyday life) are of theoretical character. If this is right, and it does seem well justified, then problems similar to those we encountered with terms "phlogiston" or "the ether" should arise with any other term, no matter how close to pure observation. Many theories agree with any given set of observations. So, there can be no guarantee that two or more theoretical terms from two or more different theories should agree with respect to observable properties and disagree with respect to the domain that is not a subject to observation (like "oxygen" and "phlogiston").

Feyerabend maintains that ontological incommensurability occurs if the concept of an object changes (cf. Jodkowski 1980b and Perovich 1991). As an example he quotes changes (cf. Feyerabend 1975, p. 264) analysed by Piaget during a child's development. This is not an example from the history of philosophy and therefore I shall present a different example – one that, I think, agrees with Feyerabend's intuition.

In Aristotelian physics substance was understood as a composition (*compositio*) of prime matter and substantial form. Substance with so-called accidents was what we might call today "an object". When Aristotelian physics and the idea of composition of matter and form were abandoned, a whole world in some sense disappeared. None of the objects of Aristotle's physics exists nowadays. A causalist might as well insist that things like tables, trees, or people still exist and that respective names still refer to them and that what changed was merely an opinion concerning the internal nature of objects. But after this revolution, surely, no substances exist and neither does prime matter nor substantial form. And terms of Aristotelian physics were not "house" or "chair" but "substance," "matter," and "form" (and a couple of others).

Let me quote, following Kuhn, two examples of observable concepts that must raise some difficulties for proponents of the causal theory of reference. The first is the reference of "water" (this example was first given by Putnam and later commented on by Kuhn) (cf. Kuhn 1986). Putnam maintained that references of "water" in 1750 and in 1950 had been the same and referred to H_2O, although chemists' views (knowledge?) had radically changed meanwhile[6]. Knowledge, in Putnam's opinion, plays no role whatsoever in determining reference. Kuhn, however, claims that this opinion of Putnam's is merely an illusion. Around 1780 a revolution in chemistry occurred. Before the date it was believed that states of aggregation (in the case of water – liquid, steam, and ice) define basic distinctions among chemical species. In 1750 only liquid was held to be water; ice and steam were not called water.

> Not until the 1780's ... was the taxonomy of chemistry transformed so that a chemical species might exist in all three states of aggregation. After that, the distinction between solids, liquids, and gases became physical, not chemical (Kuhn 1986).

The concept of internal structure or nature plays an essential role in defining substance terms. This structure (nature) is responsible for observable properties and it is to be shared by other, presently not observed, specimens of substance (e.g. water). But what observable properties ought to be considered depends strongly on the theories one accepts. Therefore, before 1780 the name "water" was applied only to some kind of liquid and after the date also to some kind of gas and to some kind of solid. The effect of the revolution was a different

---

[6] To be more precise, Putnam considered two cases of the reference of the word "water" in those years. The first regarded the Earth and the other the Twin Earth. This detail, however, is irrelevant to my discussion and therefore I leave it out.

"segmentation" of chemical reality. Some species (like ice and steam) simply disappeared.

This example may not be too convincing because a proponent of the causal theory of reference could point out that what changed was merely scientific opinion regarding the nature of water and therefore the extension of the name had changed.

The second example is much more convincing. It involves the name "planet," which concerned in Ptolemaic astronomy the so-called wandering stars, or the Moon, the Sun, Mercury, Venus, Mars, Jupiter, and Saturn. These bodies differed from the rest of stars (called the fixed stars) in that they changed their position with respect one another and to the fixed stars, as well. After Copernican astronomy was accepted, the Moon and the Sun "ceased to be" planets (shortly after that, the Sun "became" a star).

According to the causal theory of reference, the reference of the term "planet" in pre-Copernican astronomy was any celestial object whose nature involved movement with respect to the fixed stars. After the Copernican revolution, the Sun and the Moon did not change that "habit" of theirs yet they were not called planets anymore. In this change of reference the theory was decisive. Astronomical reality was "segmented" anew, according to a different theory. It ceased to be the case that the planet was every body that moved with respect to the fixed stars. Although all planets of Copernican astronomy still behaved in this way, this was not held to constitute their nature; it was rather their revolving around the Sun. The Moon was classified under a separate category of satellites (bodies revolving around the Earth), while the Sun was classified first as the motionless center of the Universe and, later, as a star. All these properties, and therefore references of terms as well, were determined on the basis of theory. A change of theory caused a change of reference of its terms. Obviously, not every change of theory is followed by a change of references.

Both examples show that there are moments in the history of science in which the hitherto obtaining articulation of the world (or of its fragment explored by a theory) changes when a new theory forces us to distinguish new objects, species, substances, etc. One can suppose this forcing consists of abandoning views on what was thought to have been the nature of model cases, with the help of which a given term was first introduced into a language. These natures either do not exist nowadays (phlogiston, the ether), or do exist but are of little importance to present-day science (water, planet). Such a change causes incommensurability of the antecedent picture of the world with the subsequent one. Proponents of the causal theory of reference thought that reference was guaranteed in their analyses of examples like gold or tiger, whose "theories" have not undergone any revolutionary changes for thousands of years. Of course, there were changes in our knowledge of tigers and gold but they were not revolutionary changes[7]. If a basic theory remains unchanged then the causal doctrine of reference does guarantee continuity of reference. But it is possible to show even for names such as "tiger" or "gold" what a revolutionary change would consist of. It would take place, I think, if some white man learned to speak and think in a language that categorizes reality in a different way, e.g. the language of Hopi or another similar language.

The hope cherished by proponents of the causal theory of reference that names are associated with objects directly, without the help of descriptions, seems illusive not only in the case of theoretical terms but even in the case of proper names. For when I decided to christen my child (I am writing about something I know from my personal experience) I did not do that at random. A telephone from the hospital that a baby was born was not enough. Information on the baby's sex was essential. The theory concerning a baby's sex is not very complicated but it is not obvious, either. My three-year-old daughter, Martha, believes that what distinguishes a boy from a girl is pants or a skirt, respectively, "but the best thing to do is just to ask them" (simply!). Moreover, my knowledge about my language plays its role, too. I must know what terms are suitable for names and which are suitable for boys and which are suitable for girls. Kripke totally rejected the role of descriptions in assigning a name to an object because a description would then describe what is necessary in its object but, as far as we know, Nixon's winning the 1968 elections or Caesar's crossing the Rubicon are not necessary. Kripke is right here but, nevertheless, the necessity that both of them should be people or, at least human-like creatures, seems acceptable[8].

---

7 After all, Kuhn doubts whether this is true with regard to gold because in that case a theory telling what degree of contamination is permissible for a sample to be regarded as gold is essential (cf. Kuhn 1986).

8 A similar argument is advanced by Ingvar Johansson (1986). (I am grateful to Dr. Zbysław Muszynski for bringing that paper to my attention.)

5. The causal theory of reference provides no explanation in what sense competing theories of what actually does not exist (e.g. the ether) regarded nevertheless the same thing

As I mentioned above, the theory of reference, which we discussed here, guarantees the continuity of reference (provided that a basic theory had not undergone a revolutionary change, i.e. during the period between revolutions). In this section, I would like to point to another of its deficiencies, which is evident if we try to apply it to guarantee continuity throughout revolutions. We know that the ether does not exist. Before the Special Theory of Relativity, however, which eliminated the belief in its existence, scientists advanced a number of competing theories about the ether. They argued about it. They pronounced mutually contradicting opinions about the same subject: the nature of the ether. The subject of these contradictory theories was the same thing: the ether. What does this statement mean, since we know that the ether is not real? What was the subject of the debates of the proponents of the ether?

The only answer the causal theory of reference can give is that those scientists talked about nothing or, in other words, about an empty set. In this sense they talked about the same thing. Undoubtedly, in some, maybe even in the fundamental sense, this answer is right since the ether is not, real. But there is another sense to our saying that scientists argued about the same object. The subject of these discussions was something other than the subject of analogous arguments of scientists investigating phlogiston, which is not as well. If we accepted the answer advanced by the causal theory of reference as the only possibility, we would then have to accept that both proponents of competing approaches to phlogiston and proponents of competing approaches to the ether talked about the same because they talked about the empty set. Presenting the issue in this way rules out, however, one sense of talking that proponents of phlogiston and proponents of the ether talked and argued about two different (although nonexistent) objects. No wonder that Nola was unable to comprehend Kuhn's remark, to the effect that Kuhn realized the existence of this sense[9]. Even incommensurable theories concern, in some sense, the same thing since they would otherwise be incapable of competing with one another.

---

[9] See note 5, in which I mention Nola's helplessly inept comment on Kuhn's words that Einstein's theory can only be accepted if we acknowledge that Newton's theory is wrong.

Preserving this sense requires a solution different from the one allowed by the causal theory of reference. I presented such a solution in my most recent book (I made use of suggestions set forth by the American philosopher, T. L. Short; cf. Jodkowski 1990, pp. 426-432).

6. Conclusion

In my most recent book I neglected a wider exposition of the causal theory of reference for (I still think this) contrary to gaudy declarations of some of its proponents, it does not guarantee to eliminate ontological incommensurability. When I was writing my book I did not, however, pay attention to a limited advantage of this theory. It guarantees a constancy of reference between scientific revolutions, even in spite of changes in scientific opinions. After a scientific revolution the world is "populated" by new objects, whose names are introduced into a language by means of the initiating causal link and which are transmitted to other members of language communities by means of language labour division. This theory can become a good theoretical foundation for Kuhn's and Feyerabend's views. For them changes of ontology take place less often than changes of scientific views. It also agrees with Kuhn's belief that students get to know the theory and the world simultaneously not only from the textbooks and lectures but also, maybe even chiefly, during practical training (in solving problems, attending lab classes, etc.).

However, one has to pay attention to the fact that some version of the descriptivist doctrine also guarantees such constancy – if a change of ontology is decided not by just any change of descriptions but by a sufficient number of them, or – in a different version – if what undergoes a change is the part of a description that describes necessary aspects of the signified object.

T. Kotarbiński Pedagogic University
Al. Wojska Polskiego 69
65-762 Zielona Góra
Poland

REFERENCES

Davidson, D. (1973-1974). On the Very Idea of a Conceptual Scheme. *Proceedings of the American Philosophical Association* **47**, 5-20.

Feyerabend, P.K. (1962). Explanation, Reduction and Empiricism. In: H. Feigl and G. Maxwell (Eds.), *Scientific Explanation, Space, & Time. Minnesota Studies in the Philosophy of Science* Vol. III. Minneapolis: University of Minnesota Press, pp. 28-97.

Feyerabend, P.K. (1965a). Problems of Empiricism. In: R.G. Colodny (Ed.), *Beyond the Edge of Certainty. Essays in Contemporary Science and Philosophy, University of Pittsburgh Series in the Philosophy of Science* vol. II. Englewood Cliffs, New Jersey: Prentice Hall, pp. 145-260.

Feyerabend, P.K. (1965b). On the "Meaning" of Scientific Terms. *The Journal of Philosophy* **62**, 97-103.

Feyerabend, P.K. (1975). *Against Method. Outline of an Anarchistic Theory of Knowledge.* London: New Left Books.

Feyerabend, P.K. (1981). *Realism, Rationalism and Scientific Method. Philosophical Papers.* Vol. 1, Cambridge-London-New York etc.: Cambridge University Press.

Jodkowski, K. (1980a). Spór o kryterium teoretyczności pojęć [The Debate over Concepts' Theoreticity Criterion]. *Studia Filozoficzne* **3**, 59-77.

Jodkowski, K. (1980b). O dwu rodzajach niewspółmierności interteoretycznej w ujęciu Paula K. Feyerabenda [On Two Kinds of Intertheoretic Incommesurability in Paul K. Feyerabend's Writings]. *Studia Filozoficzne* **7**, 79-91.

Jodkowski, K. (1981). Milczące funkcjonowanie paradygmatu [Paradigm's Tacit Functions]. *Studia Filozoficzne* **1**, 53-65.

Jodkowski, K. (1983). Problem uteoretyzowania faktów naukowych [Problem of Theory-Ladenness of Scientific Facts]. *Zagadnienia Naukoznawstwa* **19**, Issue 4, 419-435.

Jodkowski, K. (1984). *Teza o niewspółmierności w ujęciu Thomasa S. Kuhna i Paula K. Feyerabenda* [Incommensurability Thesis in Thomas S. Kuhn's and Paul K. Feyerabend's View]. *Realizm. Racjonalność. Relatywizm.* Vol. 1. Lublin: Wydawnictwo UMCS.

Jodkowski, K. (1990). *Wspólnoty uczonych, paradygmaty i rewolucje naukowe* [Scientific Communities, Paradigms and Scientific Revolutions]. *Realizm. Racjonalność. Relatywizm.* Vol. 22. Lublin: Wydawnictwo UMCS.

Johansson, I. (1986). Levels of Intension and Theories of Reference. *Theoria* **52**, 1-15.

Kitcher, Ph. (1978). Theories, Theorists, and Theoretical Change. *Philosophical Review* **8**, 519-547.

Kuhn, T.S. (1970a). *The Structure of Scientific Revolutions.* International Encyclopedia of Unified Science. Vol. II, No. 2, Second Edition, Enlarged. Chicago: University of Chicago Press.

Kuhn, T.S. (1970b). Postscript – 1969. In: Kuhn 1970a, pp. 174-210.

Kuhn, T.S. (1970c). Reflections On My Critics. In: I. Lakatos and A.E. Musgrave (Eds.), *Criticism and the Growth of Knowledge.* Cambridge: Cambridge University Press, pp. 231-278.

Kuhn, T.S. (1986). Possible Worlds in History of Science. Nobel Symposium – August 1986, typescript [Polish translation: Możliwe światy w historii nauki (Sympozjum noblowskie – sierpień 1986), *Literatura na Świecie* 1991, Issue 5 (238), pp. 120-154].

Kuhn, T.S. (1987). What Are Scientific Revolutions? In: L. Krüger, L.J. Daston, and M. Heidelberger (Eds.), *The Probabilistic Revolution*, Vol. 1: *Ideas in History*, A Bradford Book. Cambridge, Ma. etc.: MIT Press.

Nola, R. (1980). 'Paradigms Lost, or the World Regained' – An Excursion into Realism and Idealism in Science. *Synthese* **45**, 317-350.

Perovich, A. (1991). Incommensurability, Its Varieties and Its Ontological Consequences. In: G. Munévar (Ed.), *Beyond Reason. Essays on the Philosophy of Paul Feyerabend*. Dordrecht-Boston-London: Kluwer Academic Publishers, pp. 313-328.

Putnam, H. (1981). *Reason, Truth and History*. Cambridge-London-New York etc.: Cambridge University Press.

Putnam, H. (1983). Reference and Truth. In: H. Putnam, *Realism and Reason. Philosophical Papers*. Vol. 3, Cambridge-London-New York etc.: Cambridge University Press, pp. 69-86.

Quine, W.V.O. (1960). *Word and Object*. Cambridge: MIT Press.

Skalski, T. (1991). Teza o niewspółmierności teorii naukowych a nowa teoria oznaczania [Thesis of Incommensurability of Scientific Theories and the New Theory of Reference]. In: Z. Muszyński (Ed.), *Język, znaczenie, rozumienie i relatywizm* [Language, Meaning, Understanding, and Relativism]. *Biblioteka Myśli Semiotycznej*, Vol. 6. Warszawa: Wydawnictwo UMCS, pp. 47-63.

Poznań Studies in the Philosophy of the Sciences and the Humanities
2000, Vol. 72, pp. 153-170

Wolfgang Balzer

ON APPROXIMATIVE REDUCTION

ABSTRACT. A general notion of approximative reduction is defined in formal detail and applied to the Kepler-Newton example. The general definition continues an approach by (Mayer 1981a, b) in which the two features of approximation and reduction are clearly separated. The treatment of the example generalizes and improves previous accounts in which time intervals were restricted to compact intervals. In the present paper the general case of open time intervals is admitted.

0. Introduction

The aim of this paper is to describe as clearly as possible a concept of approximative reduction which generalizes the notion proposed in (Mayr 1981a, b), and to apply it in full detail to the example of the comparison of Kepler's and Newton's theories of planetary motion. I will not try to argue for this notion in a systematic way, by comparing it with alternative proposals. This would require too much space because most alternative concepts as found in (Scheibe 1973), (Krajewski 1977), (Ludwig 1978), (Moulines 1980), (Pearce and Rantala 1983, 1984), (Ehlers 1986), (Balzer, Moulines and Sneed 1987), (Rott 1991) are either described in a semi-formal way only, so that comparison would involve a substantial amount of reconstruction, or are technically very involved so that presentation of the technical apparatus would consume much energy and space. In the course of my exposition some comparative remarks will be made, however. Those formal details which I think are not central to the understanding, as well as the proofs of the theorems, are stated in the appendix.

The main achievements of the present notion of approximative reduction in my view are the following. First, it takes seriously the reductive component of approximative reduction being obtained from reduction proper by adding some kind of approximation. In some approaches, e.g. (Ludwig 1978) or (Moulines 1980), this reductive component is hardly visible. Second, the notion is very flexible in that it is not committed to a particular notion of reduction. The 'approximation-component' of the present notion may be added to any given notion of reduction. Third, the notion clearly isolates the area of counterfactual 'content' of the reduced theory while at the same time keeping this area within minial bounds. In this respect it represents a serious alternative to accounts stressing revision of theory instead of continuity, like for instance (Rott 1991). Fourth, the notion is basically very simple, and may be expressed by correspondingly simple pictures.

1. Empirical Theories

Approximative reduction obtains among *empirical theories*, therefore a precise account of approximative reduction presupposes an equally precise account of what is an empirical theory. I will use a variant of the structuralist model of empirical theories as the conceptual frame and background on which the notion of approximative reduction is defined. According to this model[1] an *empirical theory* T has the following components

$$T = \langle M_p, M, M_{pp}, U, I \rangle$$

where, for a fixed structure type[2] τ and for STR(τ) denoting the class of all structures of type τ:

$\emptyset \neq M_p \subseteq STR(\tau)$
$\emptyset \neq M \subseteq M_p$
M_{pp} is the class of all substructures of elements of M_p
U is a uniformity[3] on M_{pp}
$I \subseteq M_{pp}$.

In addition, the class M_p is characterized by axioms such that each axiom for M_p contains only symbols for at most one of the relations occurring in members of M_p, while M is characterized by axioms among which there is at least one *cluster law*, i.e. a sentence containing two or more relational symbols which cannot be written as a conjunction of sentences each containing only one such symbol. The notion of

---

1 Compare (Balzer *et al.* 1987). The more general structure of partial potential models to be used in the following is described in (Balzer 1985).

2 See appendix.

3 See appendix.

substructure is defined as follows. If $x \in M_p$ then y is a *substructure* of x iff $y \in STR(\tau)$, and each component of y is a subset of the corresponding component of x.

The elements of M_p are called *potential models*, those of M are called *models*, and those in M_{pp} *partial potential models*. The structure type τ is fixed in such a way that each potential model (and therefore also each model and each partial potential model) has the form

$$\langle D_1, \ldots, D_k, A_1, \ldots, A_m, R_1, \ldots R_n \rangle$$

where $D_1, \ldots, D_k$ are sets of *objects*, $A_1, \ldots, A_m$ are sets of *mathematical* objects, and $R_1, \ldots, R_n$ are relations and functions of types given by τ over the sets $D_1, \ldots, D_k, A_1, \ldots, A_m$. Following (Ludwig 1978), $D_1, \ldots, D_k$ are called *principal base sets*, $A_1, \ldots, A_m$ are called *auxiliary base sets*, $D_1, \ldots, D_k$, $A_1, \ldots, A_m$ are called *base sets*. The auxiliary base sets are sets of mathematical objects in their standard interpretation, and the axioms characterizing these sets are not included among the axioms characterizing the models.[4] The structures fixed in this way can be regarded in the sense of model theory as structures for a many-sorted (k sorts) higher-order formal language with a fixed mathematical part, consisting of $A_1, \ldots, A_m$, in standard interpretation. Our notion of substructure is more general than that of model theory. If x has the form $\langle D_1, \ldots, D_k, A_1, \ldots, A_m, R_1, \ldots, R_n \rangle$ then each substructure y of x has the same form, $y = \langle D_1', \ldots, D_k', A_1', \ldots, A_m', R_1', \ldots, R_n' \rangle$. Some components of y, however, may be empty[5]. In particular, the substructures may be finite, and therefore may be taken as representatives of data[6]. Note that M_{pp} is a very comprehensive class, and contains both M_p and M as subclasses. Therefore, introducing a uniformity U on M_{pp} 'automatically' yields uniformities on M_p and M by means of the well known notion of a uniformity *induced* by a subset of a uniform space.[7] Also note that

---

[4] This form of structures as well as the way of treating 'underlying' mathematical theories is taken from (Bourbaki 1968).

[5] This requires that we define a structure in general such that its base sets may be empty. I exclude constants from occurring among the relations $R_1, \ldots, R_n$ but only in the present paper, and for reasons of simplicity. The definitions may be easily extended to cover constants, but only at the cost of a certain clumsiness.

[6] This feature will not be essential here, but is essential for many other meta-theoretical issues.

[7] In particular, this saves me the construction of *empirical* uniformities out of uniformities on the potential models in the way of (Balzer *et al.* 1987), Chap. 7.

M_{pp} is explicitly defined in terms of M_p, and is included in the list of components of an empirical theory only for reasons of simplicity.

Basically, an empirical theory is built around two central items. On the one hand, there is the class of models corresponding to a set of axioms for the theory. On the other hand, there is the set *I*, the members of which I call *intended systems*[8]. Intended systems are set theoretic representations of the *real* systems to which the group of scientists working with theory *T* wants or intends to apply *T*. The determination of *I* involves irreduceably pragmatic features, and has been discussed at length elsewhere[9]. The other two items, M_p and M_{pp} are needed to create conceptual possibilities extending those of proper models, while the component *U* is needed for modelling the match between models and intended systems or the *process of application* of models to intended systems. Usually, no full match can be achieved, rather the match is approximative, i.e. it obtains only to a certain, given degree. This degree is taken from the uniformity *U*. Formally, *U* is a set of *degrees of approximation* or *blurs u*, each blur *u* being a subset of $M_{pp} \times M_{pp}$, and propositions of the form '$\langle x, y\rangle \in u$' express that the two partial potential models x, y are similar in degree *u*. Often, the blurs can be defined corresponding to positive real numbers ($u = u_{\in}$), and '$\langle x, y\rangle \in u_\varepsilon$' can be read as 'the distance between structures *x* and *y* is ε'.

With the help of an empirical theory we may formulate the proposition which Sneed has called the *empirical claim* of the theory. This is the proposition that each intended system can approximatively be embedded in a proper model.[10] The degree of approximation has to be fixed by some blur $u \in U$. Other formulations of empirical claims are possible, in particular various extensions comprising several empirical theories in the above sense have been studied under the labels of *theory nets and theory holons*[11].

---

[8] In previous publications the label 'intended applications', introduced in (Sneed, 1971), was used, which created quite some misunderstandings. I think the present label is less ambiguous.

[9] See, for instance, (Balzer *et al.* 1987), especially Chaps. 2 and 8.

[10] My favourite way of making this precise which differs from other structuralist accounts is this: $\forall x \in I\, \exists y\, \exists z(\langle x,z\rangle \in u \wedge z \sqsubseteq y \wedge y \in M)$.

[11] See, for instance, (Balzer *et al.* 1987).

2. The Topological Closure of an Empirical Theory

The notion of topological closure is a generalization of the relation between the set of rational numbers and that of real numbers, the latter being the topological closure of the former. The need for closure comes from the fact that in the set of rational numbers we find sequences of numbers which behave like convergent sequences but are not convergent *in* the set of rational numbers: their limit is not rational. For instance, we may define a sequence of rational numbers converging to the number $\sqrt{2}$ which is not rational.

Sequences of that kind are called Cauchy-sequences. Completion of a set consists in constructing the 'smallest' superset such that all Cauchy-sequences in the superset really converge to some element of that set. The idea of construction is simple. Take the elements of the superset to be equivalence classes of Cauchy-sequences from the original set, and identify each element α of the original set with the constant Cauchy-sequence $(\alpha_i)_{i \in N}$ with $\alpha_i = \alpha$. Under this construction, in the complete superset each Cauchy-sequence converges against 'itself'.

This idea may be generalized to uniform spaces. In the present context I consider a uniformity U over the class M_{pp} of partial potential models. In such a uniform space $\langle M_{pp}, U \rangle$ the notion of a Cauchy-sequence is well defined. A Cauchy-sequence is a sequence $(x_i)_{i \in N}$ such that, for all $u \in U$ there is some $n_0 \in \aleph$ such that for all $i, j > n_0$: $\langle x_i, x_j \rangle \in u$. That is, the structures in the sequence get arbitrarily close to each other when their index increases. In approximative reduction this notion is applied to the class of models of the reducing theory. I consider Cauchy-sequences of proper models of the reducing theory. These sequences do not converge in general in the sense that their 'limit' is not a proper model. In order to open up the possibility of talking about limits in these cases we have to complete the class of models along the lines indicated. The completion is constructed as the set of all equivalence classes of Cauchy-sequences.

In general, the completion of a uniform space affords reference to Cauchy-filters rather than Cauchy-sequences. I avoid this complication[12] by restricting myself to uniformities which 'come from' metric spaces. In applications to examples from empirical science usually uniformities of this more restricted kind are met.

---

[12] Actually, the difference is less in complication than in abstractness. The notion of equivalence is more intuitive for sequences than for filters.

I say that a uniform space[13] $\langle X, U\rangle$ is *separated* iff $\cap U = \Delta(X)$ where $\Delta(X)$ denotes the *diagonal* in X, i.e. the set $\{\langle x, x\rangle / x\in X\}$. A separated space I call *special* iff, in addition, it satisfies the following requirement of denumerability: there is a sequence $(u_i)_{i\in N}$ in U such that for all $u\in\aleph$ there is $i\in\aleph$ with $u_i \subseteq u$[14].

If $\langle X, U\rangle$ is a special uniform space, then two Cauchy-sequences (x_i), (y_i) in X are called *equivalent* iff, for all $u\in U$ there is $i_0\in\aleph$ such that for all $i \geq i_0$: $\langle x_i, y_i\rangle \in u$. It is easy to show that this definition, in fact, yields a relation of equivalence. The equivalence class of sequence (x_i) will be denoted by $[x_i]$, and the set of equivalence classes of sequences from a subset $Y \subseteq X$ by $[Y]$.

A special uniform space $\langle X, U\rangle$ is called *complete* iff every Cauchy-sequence in X converges in X. A well known result from topology now states that every uniform space, and in particular every special uniform space can be completed as stated in the following theorem.

> **Completion Theorem** If $\langle X, U\rangle$ is a special uniform space then up to isomorphism there is a complete, special uniform space $\langle [X],[U]\rangle$ such that X is isomorphic to a dense subset from $[X]$. The elements of $[X]$ are equivalence classes $[x_i]$ where (x_i) is a Cauchy-sequence in X[15].

As in an empirical theory the class M_{pp} comprises both M and M_p it is sufficient to introduce uniformity and completion on M_{pp}. This induces uniformities and completion on the subclasses M and M_p by means of a standard definition. If $\langle X, U\rangle$ is a uniform space, and $Y \subseteq X$ then the uniform space *induced by* $\langle X, U\rangle$ on Y consists of all blurs v of the form $v = u \cap (X \times X)$ for $u \in U$.

With these preparations I now can define the completion of an empirical theory. If $T = \langle M_p, M, M_{pp}, U, I\rangle$ is an empirical theory and $\langle X, U\rangle$ is special then the *completion* $[T]$ of T is defined by

$$[T] = \langle [M_p], [M], [M_{pp}], [U], [I]\rangle$$

---

13 The axioms for such spaces are stated in the appendix.

14 Separation plus this kind of denumerability is equivalent to the fact that the uniformity can be defined by a metric on X, see (Schubert 1964, p. 119). These two requirements are, however, easier to handle in applications. I am indebted to Thomas Mormann here.

15 For a proof see, for instance, (Schubert 1964, Sec. II.3). Our addition about the requirement of denumerability is straightforward.

where

1) $\langle [M_{pp}], [U]$ is the completion of $\langle M_{pp}, U \rangle$
2) $[I] = \{[x_i] / \forall i, j \in \aleph (x_i = x_j \in I)\}$

According to the previous definitions, $[M]$ is the class of all equivalence classes $[x_i]$ where (x_i) is a Cauchy-sequence in M ($M \subseteq M_{pp}$), and analogously for $[M_p]$. Requirement 2) says that, in the completed theory, the intended systems consist of equivalence classes of constant sequences in I. Each constant sequence is a Cauchy-sequence, so the previous definitions can be applied. The uniform spaces induced by $\langle [M_{pp}], [U] \rangle$ on $[M]$ and $[M_p]$ I denote by

$$\langle [M], [U_M] \rangle \text{ and } \langle [M_p], [U_{Mp}] \rangle$$

in the following.

3. The Notion of Approximative Reduction

Following (Mayr 1981), the basic idea of approximative reduction is simple. Suppose we have two given empirical theories T, T^* and a general definition of what it means that one empirical theory strictly reduces to another one. Then T is said to be *approximatively reduceable* to T^* iff T strictly reduces to the completion of T^*.

As already noted, this definition is completely neutral with respect to the notion of strict reduction. It applies whenever we have chosen some notion of strict reduction as holding between empirical theories in the format described above. Even without reference to a particular such notion of strict reduction the idea of *approximative* reduction may be informally grasped. The initial situation for approximative reduction is that there is no strict reduction relation among the two theories to be compared, but a kind of reduction seems possible when some limit is taken. The present notion is committed to go to the limit on the side of the reducing theory – which in historical perspective is usually the 'new' and 'better' one. Though *a priori* I cannot rule out cases in which the limit should be taken on the reduced side (the 'old' theory) I neglect this possibility for two reasons. First, in all cases which have been studied up to now the limit is taken on the reducing side. This is so in the transitions from Kepler's to Newton's planetary theory, from classical to relativistic physics, and from classical to quantum physics[16]. Second, the reducing theory usually has a richer conceptual apparatus including further parameters which in the reduced theory are simply not present. It is in such additional conceptual areas that limits can be, and are, taken.

Considering a given potential model we may concentrate on one of its components for which it makes sense to go to some limit and see what happens when this component is replaced by successively different values approaching the limit. Simple examples are given by a component just being a numerical constant (like the velocity of light or Planck's constant). If we replace the component by successive values we obtain a sequence of corresponding potential models which all look like the initial one, and differ from it only in the component under consideration. In this way we can generalize talk about one component's going to a limit to simply talking about a sequence of potential models going to some limit. In the present frame this is possible on the very general level of the uniformity which is part of the theory: 'going to the limit' can be represented by sequences of (potential or partial potential) models which converge.

In going to a limit two levels may be distinguished. First, the limit may be taken purely conceptually without worrying about implications for the laws. In my frame this amounts to consideration of a sequence of potential or partial potential models some components of which go to a limit while the other components remain unchanged. On a second level, we may ask whether such sequences are compatible with the laws obtaining in the theory. A sequence of potential models (x_i) thus may be said to be compatible with the theory's laws iff all x_i, in fact are proper models, and the sequence may be said to be compatible with the theory's laws *in the limit* iff its limit also is a proper model (provided the sequence has a limit in the class of potential models). Now the crucial observation about examples from the sciences is that in going to the limit we leave the domain of proper models. If we start with a model and let the crucial parameter go to infinity in a sequence of values, we obtain a sequence of corresponding potential models -one for each value of the parameter- which in a certain sense converges to some limit structure which again is a potential model. The crucial point is that, for each finite value of the parameter, the corresponding structure in the sequence is still a proper model while for the limit value the structure is no longer a model.

---

16 Very crudely, in the first case the sun's mass, in the second case the velocity of light, and in the third case the inverse of Planck's constant are going to infinity.

This is the point where the completion of the reducing theory comes in. Even though a sequence of models obtained by letting a parameter go to infinity does not converge to a model, it does converge to a model in the completion of the theory. This is so for purely conceptual grounds. As the sequence converges to some potential model in the original theory it must be a Cauchy-sequence. Each member of the sequence is a proper model, so its 'completed image'[17] belongs to the completion $\langle [M], [U_M] \rangle$. In the completed theory the sequence of these images is a Cauchy-sequence and therefore converges by virtue of the theory being complete. So there is some limit, and this limit is, by definition, a member of $[M]$, i.e. a model of the completed theory. This situation is depicted in Figure 1.

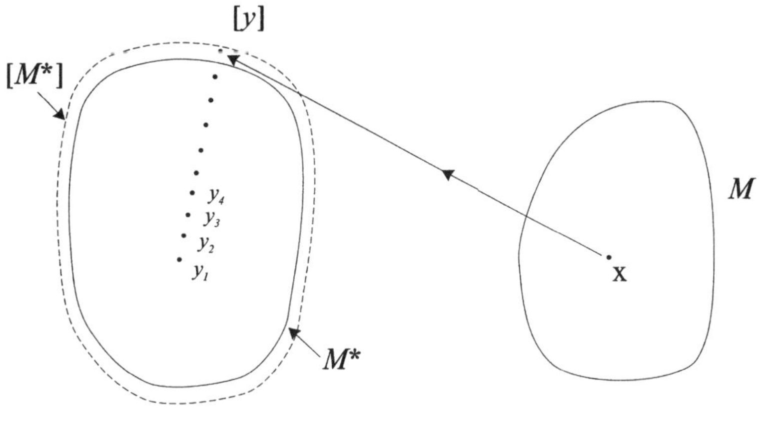

The model classes M and M^* of the two theories under comparison are drawn as sets of points while the completed set of models $[M^*]$ is indicated by the dotted line. The 'completed' models are indicated by square brackets. The arrow indicates that the model x under a given strict reduction relation corresponds to the 'limit-model' y.

So, passing over to the completion of the reducing theory we have the possibility to form limits of sequences of proper models such that these limits are again models. The completed theory then may enter into a strict reduction relation with the reduced theory.

[17] That is, the equivalence class of the constant sequence formed out of this model.

In order to make this notion of approximative reduction applicable I have to choose a notion of strict reduction. In the light of extensive recent discussion[18] my preferred notion of strict reduction is this.

D1 Let $T = \langle M_p, M, M_{pp}, U, I\rangle$ and $T^* = \langle M_p^*, M^*, M_{pp}^*, U^*, I^*\rangle$ be empirical theories. I say that T by means of ρ is *directly reduced to* T^* iff

1) $\rho \subseteq M_{pp} \times M_{pp}^*$
2) $\forall x \in M_p \exists x^* \in M_p^* (\langle x, x^*\rangle \in \rho)$
3) $\forall x \in M \exists x^* \in M^* (\langle x, x^*\rangle \in \rho)$
4) $\forall x \in I \exists x^* \in I^* (\langle x, x^*\rangle \in \rho)$

By 1), ρ relates partial potential models, and among these it is the models and potential models which 'really count'. By 2) ρ is defined for every potential model on the reduced side[19]. This is intended to express that ρ has to be defined merely by reference to the terms of T. Condition 3) says that each model of the 'old' theory can be 'embedded' via ρ into a corresponding model of the 'new' theory. Conditions 4) extends this requirement to intended systems. Each intended system of the 'old' theory 'is' (in the sense of being ρ correlated to) an intended system of the 'new' theory.

I have dropped the condition of uniqueness of ρ present in many accounts[20] because I think it is too strong to apply to all real cases[21]. I have replaced the original Sneedian condition

5) $\forall x, x^* (\langle x, x^*\rangle \in \rho \wedge x^* \in M^* \Rightarrow x \in M)$

by 3) above because 5) allows to shift the content of the condition of 'derivability' of axioms to the definition of the ρ-relation[22]. Moreover, 3) has the more constructive appeal of an embedding[23]. Note that both conditions, 3) and 5) may be regarded as expressing the requirement that the axioms of the reduced theory are derivable from those of the reducing theory 'after translation', so reference to derivability does not yield argu-

---

18 See (Mormann 1988), (Pearce 1987), (Rott 1991).

19 We even might extend this requirement to the full level of partial potential models.

20 For instance in (Sneed 1971) or (Pearce 1987).

21 See the examples in Chap. 6 of (Balzer *et al.* 1987) including those arising in the context of equivalence.

22 In the extreme case, ρ might be defined by adding a conjunction '$\langle x, x^*\rangle \in \rho \Rightarrow x \in M$'.

23 This condition has been proposed already in (Bourbaki 1968).

ments for or against each version[24]. Finally, I stress that the present definition of reduction differs from the original structuralist versions in that it is stated without reference to a distinction between theoretical and non-theoretical terms. The difficulty of drawing such a distinction in real-life examples has beset application of the structuralist approach. I can avoid this difficulty by means of my more general notion of partial potential models[25] which is independent of distinguishing theoretical terms. This approach also allows to state condition 4) above without extra definition.

I say that T by means of ρ is *directly approximatively reduced* to T^* iff T is directly reduced by ρ to the completion $[T^*]$ of T^*. I say that T by means of ρ is *approximatively reduced* to T^* iff there is some specialization[26] T_0^* *of* T^* such that ρ directly approximatively reduces T to T_0^*.

4. Example: Kepler-Newton

The case of approximative reduction of Kepler's theory of planetary motions to Newtonian theory of gravitation has been treated by *Scheibe* whose axiomatizations[27] of the two theories I take over. His explicit relation is restricted to two-particle systems while (Mayr 1981) extends the relation to n-particle systems, but at the price of restricting the systems to compact intervals of time. I extend Mayr's account to cover the general case of open intervals representing time.

D2 x is a *model of Kepler's theory* ($x \in M(KEP)$) iff there exist P, T, s, μ, p_0 such that $x = \langle P, T, \mathfrak{R}, \mathfrak{R}^3, s, \mu, p_0\rangle$ and

1) P is a finite, non-empty set and $p_0 \in P$
2) $T \subseteq \mathfrak{R}$ is an open interval
3) $s : P \times T \to \mathfrak{R}^3$ is twice continuously differentiable in the second argument
4) $\mu : P \to \mathfrak{R}$
5) $\mu(p_0) > 0$ and for all $p \in P, p \neq p_0 : \mu(p) = 0$

---

24 This point is discussed in detail in (Balzer *et al.* 1987), Chap. 6.

25 This notion was first introduced in (Balzer 1982).

26 $T^0 = \langle M_p^0, M^0, M_{pp}^0, U^0, I^0\rangle$ is a *specialization* of $T = \langle M_p, M, M_{pp}, U, I\rangle$ iff $M_p^0 = M_p$, $M^0 \subseteq M$, $I^0 \subseteq I$, and U^0 is finer than the uniformity induced on M^0 by U.

27 (Scheibe 1973). This is not the place to discuss the *historical* adequacy of these reconstructions.

6) For all $p, p' \in P$ and all $t \in T$:
6.1) if $p \neq p'$ then $s(p, t) \neq s(p', t)$

6.2) if $p \neq p_0$ then $\frac{1}{2}|\dot{s}(p,t)-\dot{s}(p_0,t)| - \mu(p_0)\,|s(p,t)-s(p_0,t)|^{-1} < 0$

6.3) $s(p,t) = \sum_{p' \neq p, p' \in P} \mu(p') \frac{s(p,t)-s(p',t)}{|s(p,t)-s(p',t)|^3}$

Systems satisfying 1)–5) I call *potential models* of Kepler's theory, and from their class $M_p(KEP)$ the class of partial potential models $M_{pp}(KEP)$ ist defined as described in Sec. 1. Let $KEP = \langle M_p(KEP), M(KEP), M_{pp}(KEP), U(KEP), I(KEP)\rangle$ where $U(KEP)$ is not specified here, and $I(KEP)$ is assumed to be fixed pragmatically.

Intuitively, the elements of P are the particles making up the system, p_0 denoting a distinguished particle: the 'sun'. T represents the interval of time in which the system is considered, and $\mathfrak{R}$ represents space (relative to a fixed, implicit inertial system). s is the *position function* assigning a 'position vector' to each particle at each point of time. μ is a generalized mass function which, by 5), assigns some positive mass only to the distinguished particle p_0[28]. 6.1) excludes 'colliding' particles. 6.2) and 6.3) are Kepler's laws in modern formulation. They function as follows. 6.2) assures that the system has closed solutions (excluding parabolas and hyperbolas) and 6.3) yields non-trivial equations between $p \neq p_0$ on the left, and p' on the right-hand side (the sum shrinks to one expression for $p' = p_0$). These differential equations yield ellipses as solutions.

D3 x^* is a *model of Newtonian theory of gravitation* ($x^* \in M(GRAV)$) iff there exist $P^*, T^*, \mathfrak{R}, \mathfrak{R}^3, s^*, m$ such that $x = \langle P^*, T^*, \mathfrak{R}, \mathfrak{R}^3, s^*, m\rangle$ and
1) P^* is a finite-non-empty set
2) $T^* \subseteq \mathfrak{R}$ is an open interval
3) $s^*: P \times T \rightarrow \mathfrak{R}^3$ is twice continuously differentiable in the second argument
4) $m: P \rightarrow \mathfrak{R}$ and for all $p \in P: m(p) > 0$
5) For all p, p', t, if $p \neq p'$ then $s^*(p, t) \neq s^*(p', t)$
6) For all $p \in P$ and $t \in T$:

---

[28] This way of stating μ is due to D. Mayr.

$$s(p,t) = \sum_{p' \in P, p' \neq p} m(p') \frac{s(p,t) - s(p',t)}{|s(p,t) - s(p',t)|^3}$$

Systems satisfying 1)-5) are called *potential models*, and the class of partial potential models $M_{pp}(GRAV)$ is defined by substructures as described in Sec.1. P^*, T^*, s^*, and 5) have the same interpretation as in *KEP*. *m* is the classical *mass function*, and 6) states the law of gravitation.

In order to apply the apparatus of completion I first have to introduce a uniformity on $M_{pp}(GRAV)$. The inclusion of open intervals of time here yields some complication. Intuitively, I will first 'shrink' the open interval of time to the *bounded* interval $]-\frac{\pi}{2},+\frac{\pi}{2}[$, and then work with the latter. To this end, let $f:]-\frac{\pi}{2},+\frac{\pi}{2}[\to \Re, f(x) = arctg(x)$, and let $\mathbf{m}$ be the measure induced by f and the Lebesgue measure on $]-\frac{\pi}{2},+\frac{\pi}{2}[$. As a matter of real analysis I use the following statement:

(1) $\forall \in > 0 \exists i_0 \in \aleph \forall i, j \geq i_0$: if $i < j$ then $\mathbf{m}([i, j[\cup]\frac{1}{j}, \frac{1}{i}]) < \in$

In the following definition, let λ denote the Lebesgue measure, $\|X\|$ the cardinality of the set *X*, and $X \Delta Y$ the symmetric difference of sets *X* and *Y*: $X \Delta Y = (X \cup Y) \setminus (X \cap Y)$.

D4 For $\varepsilon > 0$ $u_\varepsilon \subseteq M_{pp}(GRAV) \times M_{pp}(GRAV)$ is defined as follows. For all x, y, $\langle x, y \rangle \in u_\varepsilon$ iff

1) $x = \langle P, T, \Re, \Re^3, s, m \rangle \in M_{pp}(GRAV)$ and $y = \langle P', T', \Re, \Re^3, s', m' \rangle \in M_{pp}(GRAV)$
2) $P = P'$
3.1) $T \Delta T'$ is a borel set
3.2) $\|T \Delta T'\| < \varepsilon$, if $\|T \Delta T'\| < \infty$
$\mathbf{m}(T \Delta T') < \varepsilon$, if $\|T \Delta T'\| \geq \infty$ and $\lambda(T \Delta T') \neq 0$
4) For all $p \in P$ and $t \in T \cap T'$:
4.1) $|s(p,t) - s'(p,t)| < \varepsilon$
4.2) $|m(p) - m'(p)| < \varepsilon$

U(*GRAV*) is defined as the class of all $u \subseteq M_{pp}(GRAV) \times M_{pp}(GRAV)$ for which there is some $\varepsilon > 0$ such that $u_\varepsilon \subseteq u$. By *GRAV* I denote the theory $\langle M_p(GRAV), M(GRAV), M_{pp}(GRAV), U(GRAV), I(GRAV) \rangle$ where $I(GRAV)$ is assumed to be fixed pragmatically.

Theorem 1 $\langle M_{pp}(GRAV), U(GRAV)\rangle$ is a special uniform space.

Let $[GRAV] = \langle [M_p(GRAV)], [M(GRAV)], [M_{pp}(GRAV)], [U(GRAV)], [I(GRAV)]\rangle$ be the completion of the theory of gravitation *GRAV* just defined. In order to complete the approximative reduction of *KEP* to *GRAV* according to the previous definitions I have to specifiy a relation ρ strictly reducing *KEP* to $[GRAV]$.

D5 I define $\rho \subseteq M_{pp}(KEP) \times [M_{pp}(GRAV)]$ as follows. For all x,y: $\langle x, y\rangle \in \rho$ iff there exist $P, T, s, \mu, p_0, P_i, T_i, s_i, m_i, x_i, (i \in \aleph)$ such that $x = \langle P, T, \mathfrak{R}, \mathfrak{R}^3, s, \mu, p_0\rangle$ $y = [x_i]$, where, for $i \in \aleph$, $x_i = \langle P_i, T_i, \mathfrak{R}, \mathfrak{R}^3, s_i, m_i\rangle$ and there is an increasing sequence $(T_i^*)_{i \in N}$ of compact intervals such that for all $i \in \aleph$:
1) $P_i = P$, T_i is the interior of T_i^* and $T = \cup\{T_i^* / i \in \aleph\}$:
2) For all $p \in P$ and $t \in T$: $lim_{i \to \infty} s_i(p, t) = s(p, t)$
3) For all $p \in P : lim_{i \to \infty} m_i(p) = \mu(p)$.

Thus ρ relates Keplerian systems with sequences of gravitating systems in such a way that the particles in both systems are the same and the other components in the gravitating systems approach the respective components of the Kepler system in the limit. The compact time intervals T_i^* of the gravitating systems 'converge' (by taking their union) to the time interval T of the Kepler system (condition 1), the position functions s_i of the gravitating systems converge to that of the Kepler system (condition 2), and the masses of the gravitating systems converge to the 'Kepler-mass' μ (condition 3). In condition 2) the limit is to be taken for those indices i for which t is an element of T_i^*. Once t is in some such T_i^* it is in all succeeding intervals $T_j^*, j > i$. Condition 3) concerning the masses is crucial. As $\mu(p)$ is zero for all p different from the 'sun' p_0, 3) says that all planets' masses go to zero, while that of the 'sun' p_0 goes to the Kepler constant $\mu(p_0)$. The planets' masses going to zero here is a counterfactual condition, and it is this condition which -in the limit- gets us outside of the range of gravitational models. For according to condition 4) in the definition of the latter models, $m(p) > 0$ for all p.

Theorem 2 If D1-4 is satisfied for T = *KEP* and T^* = *GRAV* then ρ directly reduces *KEP* to [GRAV] and thus approximately reduces KEP to *GRAV*.

Appendix

D6 Let k, $m \in \aleph$. $(k + m)$-*types* σ are inductively defined. 1) For each $i \leq k + m$, $[i]$ is a $(k + m)$-type. 2) If σ_1 and σ_2 are $(k + m)$-types then so are $(\sigma_1 \times \sigma_2)$ and $pow(\sigma_1)$. τ is a *structure type* iff there exist k, $m, n, \sigma_1, \ldots, \sigma_n$ such that $\tau = \langle k, m, n, \sigma_1, \ldots, \sigma_n \rangle$, $k, m, n \in \aleph$, $k > 0$, $n > 0$, and for all $i \leq$ n: σ_i is a $(k+m)$-type.
If σ is a $(k + m)$-type, $D_1, \ldots, D_k, A_1, \ldots, A_m$ are sets, and $u = \langle D_1, \ldots, D_k, A_1, \ldots, A_m \rangle$ then $E_\sigma(u)$, the *echelon set* of type σ over u is defined by induction wrt. the definition of σ. 1) If σ has the form $[i]$ then $E_\sigma(u) = D_i$. 2) If σ has the form $(\sigma_1 \times \sigma_2)$, resp. $pow(\sigma_1)$, then $E_\sigma(D_1, \ldots, A_m)$ is set equal to $E_{\sigma_1}(u) \times E_{\sigma_2}(u)$ ('Cartesian product'), resp. $\mathbf{Po}(E_{\sigma_1}(u))$ ('power set').
For $\tau = \langle k, m, n, \sigma_1, \ldots, \sigma_n \rangle$, $x \in STR(\tau)$ iff there exist sets $D_1, \ldots, D_k$, $A_1, \ldots, A_m$, $R_1, \ldots, R_n$ such that 1) $x = \langle D_1, \ldots, D_k, A_1, \ldots, A_m, R_1, \ldots, R_n \rangle$,
2) for all $i \leq n$, $R_i \in \mathbf{Po}(E_{\sigma i}(D_1, \ldots, D_k, A_1, \ldots, A_m))$
3) $A_1, \ldots, A_m$ are sets of standard mathematical objects.

D7 U is a *uniformity* on X (or $\langle X, U \rangle$ is a *uniform space* iff 1) X is a non--empty set (or class), 2) $\emptyset \neq U \subseteq \mathbf{Po}(X \times X)$, 3) for all $u, v \in \mathbf{Po}(X \times X)$: 3.1) if $u \in U$ and $u \subseteq v$ then $v \in U$, 3.2) if $u, v \in U$ then $u \cap v \in U$, 3.3) if $u \in U$ then $\Delta(X) \subseteq u$, 3.4) if $u \in U$ then $u^{-1} \in U$, 3.5) if $u \in U$ then there is $w \in U$ such that $w \circ w \subseteq u$. Here, $\Delta(X) = \{\langle x, x \rangle / x \in X\}$, $u^{-1} = \{\langle y, x \rangle / \langle x, y \rangle \in u\}$, and $w \circ w = \{\langle x, y \rangle / \exists z(\langle x, z \rangle \in w \wedge \langle z, y \rangle \in w)\}$.

Theorem 1 $\langle M_{pp}(GRAV), U(GRAV) \rangle$ is a special uniform space.
Proof: Conditions 1), 2), 3.1), 3.3), and 3.4) of D7 are trivial.
D7-3.2: Let ε_u, ε_v be the numbers corresponding to given $u, v \in U$, and $\varepsilon = \min\{\varepsilon_u, \varepsilon_v\}$. I show that $u_\varepsilon \subseteq u_{\varepsilon_u} \cap u_{\varepsilon_v}$. If $\langle x, y \rangle \in u_\varepsilon$ then $P_x = P_y$.

Case 1: $\|T_x \Delta T_y\| < \infty$, and so, by D4-3, is $< \varepsilon$. So the corresponding condition for ε_u, $\|T_x \Delta T_y\| < \varepsilon \leq \varepsilon_u$ is satisfied, and similarly for ε_v. The same reasoning applies to the second case.
D7-3.5: For given $u \in U$ and corresponding ε such that $u_\varepsilon \subseteq u$ let $v = \{\langle x, y \rangle \in M_{pp}^2 \,/\, P_x = P_y \wedge T_x = T_y \wedge \forall t \in T_x \forall p \in P_x(|s_x(p,t) - s_y(p,t)| < \varepsilon/2 \wedge |m_x(p) - m_y(p)| < \varepsilon/2)\}$. If $\langle x, z \rangle \in v \wedge \langle z, y \rangle \in v$ then $P_x = P_z = P_y$, $T_x = T_z = T_y$ and for all $p \in P_x$, $t \in T_x$: $|s_x(p, t) - s_z(p, t)| < \varepsilon/2$ and $|s_z(p, t) - s_y(p, t)| < \varepsilon/2$. By means of the triangle inqueality this yields $|s_x(p, t) - s_y(p, t)| < \varepsilon$, for all p, t, and similarly for m. So $v \circ v \subseteq u_\varepsilon \subseteq u$.

Separation: Clearly, $\Delta(M_{pp}) \subseteq \cap U$. Conversely, let $\langle x, y\rangle \in \cap U$. This yields (2) $P_x = P_y$, by D5-1. Case 1: $\|T_x \Delta T_y\| < \infty$. As $\langle x, y\rangle \in u_\varepsilon$ for all ε, we may set $\varepsilon = 1/2$. It follows that $\|T_x \Delta T_y\| < \frac{1}{2}$, so $T_x = T_y$. Case 2: $\|T_x \Delta T_y\| = \infty$ and $T_x \Delta T_y$ is a borel set and $\lambda(T_x \Delta T_y) \neq 0$. Then, by D4-3, $\mathbf{m}(T_x \Delta T_y) < \varepsilon$, and therefore, since ε was arbitrary, $\lambda(T_x \Delta T_y) < \varepsilon$ for arbitrary ε. But this yields $\lambda(T_x \Delta T_y) = 0$, in contradiction to the description of the present case. So this case cannot obtain.
So in all cases (3) $T_x = T_y$. By D4-4: $|s(p,t) - s'(p,t)| < \varepsilon$ for arbitrary ε. This implies: (4) $s = s'$, and similarly we obtain (5) $m = m'$. But (2) to (5) yield $x = y$, so $\langle x, y\rangle \in \Delta(M_{pp})$.
Denumerability: The sequence (v_i) with $v_i = u_{1/i}$, $i \in \aleph$ is a denumerable base of the filter of neighbourhoods. For $u \in U$ there is $\in$ such that $u_\varepsilon \subseteq u$. So there is $i \in \aleph$ such that $1/i < \varepsilon$, and $u_{1/i} \subseteq u_\varepsilon \subseteq u$. □

Theorem 2 If D1-4 is satisfied then ρ directly reduces *KEP* to [*GRAV*].
Proof: Obviously, $\rho \subseteq M_{pp}(KEP) \times [M_{pp}(GRAV)]$. We show

(6) $\forall x \in M_p(KEP) \exists x^* \in [M_p(GRAV)](\langle x, x^*\rangle \in \rho)$. For given $x \in M_p(KEP)$ let T_x be the open interval occurring in x. As a matter of real analysis there exists an increasing sequence (T_i) of compact intervals, $T_i \subseteq T_x$ such that $\cup T_i = T_x$. We set $x_i = \langle P_x, T_i^0, \mathfrak{R}, \mathfrak{R}^3, s_i, m_i\rangle$ where T_i^0 is the interior of T_i, $m_i = \mu$ and s_i is the restriction of s_x to T_i^0.
(7) $\forall x \in M(KEP) \exists x^* \in [M(GRAV)](\langle x, x^*\rangle \in \rho)$. Let $x = \langle P, T, \mathfrak{R}, \mathfrak{R}^3, s, \mu, p_o\rangle \in M(KEP)$. Without restriction of generality we may assume that $T = \mathfrak{R}^+ = \{x \in \mathfrak{R} / x > 0\}$. We define (x_i) such that $x_i \in M(GRAV)$. Let (8) $P_i = P$, $T_i =]\frac{1}{i}, i[$, $\hat{T}_i = [\frac{1}{i}, i]$. So T_i is the interior of $\hat{T}_i$.
For $\bar{y} = \langle P, \hat{T}_i, \mathfrak{R}, \mathfrak{R}^3, s, m\rangle$ let $\tilde{y} = \langle P, T_i, \mathfrak{R}, \mathfrak{R}^3, s_{/T_i}, m\rangle$.
Lemma 1: If $\bar{y}$ satisfies the gravitational equations (D3-6) then $\tilde{y} \in M(GRAV)$. The proof is trivial.
Lemma 2: If $x = \langle P, \hat{T}_i, \mathfrak{R}, \mathfrak{R}^3, s, \mu, p^0\rangle$ satisfies the Kepler equations (D2- 6.2,6.3) then for each $\in > 0$ there is some $x' = \langle P, \hat{T}_i, \mathfrak{R}, \mathfrak{R}^3, s', m'\rangle \in M_{pp}(GRAV)$ such that (i) x' satisfies the gravtiational equations (ii) $\forall p \in P \forall t \in \hat{T}_i : |s(p,t) - s'(p,t)| < \varepsilon \wedge |\mu(p) - m'(p)| < \varepsilon$.
Proof: This follows from the fact that the gravitational equations are first order and satisfy a Lipschitz condition on the compact interval $\hat{T}_i$, compare (Mayr 1981).
Lemma 3: For all $i \in \aleph$ there is $x_i \in M(GRAV)$ such that
(i) $x_i = \langle P, T_i, \mathfrak{R}, \mathfrak{R}^3, s_i, m_i\rangle$

(ii) $\forall p \in P \forall t \in T_i$: $|s(p, t) - s_i(p, t)| < 1/i \wedge |\mu(p) - m_i(p)| < 1/i$.
Proof: Let $i \in \aleph$. Choose $\in$ so that $\in < 1/i$. As $x \in M(KEP)$, $\bar{x} = \langle P, \hat{T}_i, \Re, \Re^3, s_{/\hat{T}_i}, \mu, p_0 \rangle$ satisfies Kepler's equations. By Lemma 2 we obtain some $x' = \langle P, \hat{T}_i, \Re, \Re^3, s', m' \rangle$ which satisfies the gravitational equations and satisfies (9) $|s(p, t) - s'(p, t)| < 1/i \wedge |\mu(p) - m'(p)| < 1/i$. I set $m_i(p) = m'(p)$ and $s_i(p, t) = s'(p, t)$. From $T_i \subseteq \hat{T}_i$ we obtain $x_i = \langle P, T_i, \Re, \Re^3, s_i, m_i \rangle \in M(GRAV)$. Then (i) is satisfied by the definition of x_i, and (ii) holds because of (9) and the definition of x_i.
Now let (x_i) be chosen according to Lemma 3, i.e. $x_i = \langle P, T_i, \Re, \Re^3, s_i, m_i \rangle \in M(GRAV)$ and $\forall p \in P \forall t \in T_i$: $|s_i(p, t) - s(p, t)| < 1/i \wedge |\mu(p) - m_i(p)| < 1/i$.
Lemma 4: (x_i) is a Cauchy sequence in $M(GRAV)$.
Proof: a) $x_i \in M(GRAV)$ holds by construction of (x_i) and Lemma 3.
b) We show $\forall \varepsilon > 0 \, \exists i_0 \forall i, j \geq i_0$: $\langle x_i, x_j \rangle \in u_\varepsilon$.
With $f:]-\frac{\pi}{2}, +\frac{\pi}{2}[\rightarrow \Re, f(x) = arctg(x)$ and the Lebesgue measure λ, **m**: $= \lambda \circ f$ is a measure on $]-\frac{\pi}{2}, +[$ endowed with the corresponding borel sets. Furthermore, by (1) in Sec.4,
$\forall \varepsilon \exists j_0 \forall i,j \geq j_0$: $i < j \Rightarrow \mathbf{m}([i,j[\cup]1/j, 1, i]) < \varepsilon$. Now let ε be given in b). Choose i_1 such that $1/i_1 < \varepsilon/2$. Choose j_0 according to (1) and set $i_0 := min\{i_1, j_0\}$. Then (1) yields (10) $\forall i,j \geq i_0$: $i < j \Rightarrow \mathbf{m}([i,j[\cup]1/j, 1/i])$ $< \varepsilon$ and $1/i_0 < \varepsilon/2$. Let $i,j \geq i_0$ be given, and assume without loss of generality that $i < j$. By construction of x_i we have (11) $P_i = P_j$.
Furthermore, $T_i =]1/i, i[\subseteq]1/j, j[$, so $T_i \Delta T_j = T_j \backslash T_i =]1/j, 1/i] \cup [i, j[$. So $\|T_i \Delta T_j\| = \infty$, $T_i \Delta T_j$ is a borel set and $\lambda(T_i \Delta T_j) \neq 0$. This, by (1), yields (12) $\mathbf{m}(T_i \Delta T_j) = \mathbf{m}([i, j[\cup]1/j, 1/i]) < \varepsilon$. Let $p \in P_i = P_j$ and $t \in T_i = (T_i \cap T_j)$. By Lemma 3: $|s(p, t) - s_i(p, t)| < 1/i \leq 1/i_0$, so $\forall t \in T_j$: $|s(p, t) - s_j(p, t)| < 1/j < 1/i \leq 1/i_0$. As $T_i \subseteq T_j$ this yields
(13) $\forall t \in T_i$: $|s_i(p, t) - s_j(p, t)| \leq |s_i(p, t) - s(p, t)| + |s(p, t) - s_j(p, t)| < 2/i_0 < \varepsilon$.
Analogously, it is shown that (14) for all p: $|m_i(p) - m_j(p)| < \varepsilon$. But from (4)-(7) it follows that $\langle x_i, x_j \rangle \in u_\varepsilon$ which proves Lemma 4.
Lemma 5: $\forall \varepsilon > 0 \, \exists j_0 \forall r,s \geq j_0 \forall p \in P \forall t \in T_{j0}$: $|s_r(p, t) - s_s(p, t)| < \varepsilon \wedge |m_r(p) - m_s(p)| < \varepsilon$.
Proof: For given ε choose j_0 such that $1/j_0 < \varepsilon/2$. With $r, s \geq j_0$ we obtain from Lemma 3, as $T_{j_0} \subseteq T_r$, $\forall p \forall t \in T_{j_0}$: $|s(p, t) - s_r(p, t)| < 1/r \leq 1/j_0 \wedge |s(p, t) - s_s(p, t)| < 1/s \leq 1/j_0$, and from this with the triangle inequality $\forall p \forall t \in T_{j_0}$: $|s_r(p, t) - s_s(p, t)| < \varepsilon$. The second part is proven analogously.
Lemma 6: $\forall p \in P \forall t \in T$: $lim_{i \rightarrow \infty} s_i(p,t) = s(p,t)$.

Proof: Let $\bar{p} \in P$, $\bar{t} \in T$ be given, and $\varepsilon > 0$. By definition of ρ, $T = \cup T_i$, so there is some k such that $\bar{t} \in T_k$. By Lemma 5:
(15) $\exists j_0 \forall r, s \geq j_0 \forall p \in P \forall t \in T_{j_0:} |s_r(p, t) - s_s(p, t)| < \varepsilon/2$. Let $i_0 \geq max\{k, j_0\}$ be such that (16) $1/i_0 < \varepsilon/2$.
Let $i \geq i_0$ be given, $p \in P$, $t \in T_{i_0}$, $r \geq i_0$ and $j \geq i_0$. Then
(17) $|s_j(p, t) - s(p, t)| \leq |s_j(p, t) - s_r(p, t)| + |s_r(p, t) - s(p, t)|$. Here, the first part is smaller than $\varepsilon/2$ by (15). For the second part, note that by Lemma 3: $\forall t \in T_r : |s(p,t) - s_r(p,t)| < 1/r \leq 1/i_0 < \varepsilon/2$. As $i_0 \leq r$, and therefore $T_{i_0} \subseteq T_r$, the latter inequality holds for all $t \in T_i$. So the second part of (17) also is smaller than $\varepsilon/2$. Thus (17) yields $\forall t \in T_{i_0} \forall j \geq i_0 : |s_j(p, t) - s(p, t)| < \varepsilon$. As $k \leq i_0$, $T_k \subseteq T_{i_0}$, so $\forall j \geq i_0$: $\left|s_j(\bar{p},\bar{t}) - s(\bar{p},\bar{t})\right| < \varepsilon$
Lemma 7: For all $p \in P$: $lim_{i \to \infty} m_i(p) = \mu(p)$
Proof: By Lemma 5 $\exists j_0 \forall r, s \geq j_0 : |m_r - m_s| < \varepsilon/2$. Let $i_0 = j_0$ and $j \geq i_0$. If r is chosen such that $r \geq j_0$ and $1/r < \varepsilon/2$, we obtain by Lemma 3: $|m_j - \mu| \leq |m_j - m_r| + |m_r - m_u| < \varepsilon$. □

Institute for Philosophy, Logic and Philosophy of Science
University of Munich

REFERENCES

Balzer, W. (1982). Empirical Claims in Exchange Economics. In: W. Stegmueller *et al.* (Eds.), *Philosophy of Economics*. Berlin etc.: Springer, pp. 16-40.

Balzer, W. (1985). *Theorie und Messung*. Berlin etc.: Springer.

Balzer, W., Moulines, C.U. and Sneed, J.D. (1987). *An Architectonic for Science*. Dordrecht: Reidel.

Bourbaki, N. (1968). *Theory of Sets*. Paris: Hermann.

Ehlers, J. (1986). On Limit Relations between, and Approximative Explanations of Physical Theories. In: R. Barcan Marcus *et al.* (Eds.), *Proceedings of the 7th International Congress of Logic, Methodology and Philosophy of Science*. Amsterdam: North-Holland, pp. 386-403.

Krajewski, W. (1977). *Correspondence Principle and Growth of Science*. Dordrecht: Reidel.

Ludwig, G. (1978, 2nd ed. 1990). *Die Grundstrukturen einer physikalischen Theorie*. Berlin etc.: Springer.

Mayr, D. (1981a). Approximative Reduction by Completion of Empirical Uniformities. In: A. Hartkaemper and H.-J. Schmidt (Eds.), *Structure and Approximation in Physical Theories*. New York: Plenum, 55-70.

Mayr, D. (1981b). Investigations of the Concept of Reduction II. *Erkenntnis* **16**, 109-129.

Mormann, T. (1988). Structuralist Reduction Concepts as Structure Preserving Maps. *Synthese* **77**, 215-250.

Moulines, C.U. (1980). Intertheoretic Approximation: The Kepler-Newton Case. *Synthese* **45**, 387-412.

Pearce, D. and Rantala, V. (1983). Correspondence as an Intertheory Relation. *Studia Logica* **42**, 363-371.

Pearce, D. and Rantala, V. (1984). A Logical Study of the Correspondence Relation. *Journal of Philosophical Logic* **13**, 47-84.

Pearce, D. (1987). *Roads to Commensurability.* Dordrecht: Reidel.

Rott, H. (1991). *Reduktion und Revision.* Frankfurt/M.: Peter Lang.

Scheibe, E. (1973). Die Erklaerung der Keplerschen Gesetze durch Newtons Gravitationsgesetz. In: E. Scheibe and G. Suessmann (Eds.), *Einheit und Vielheit - Festschrift fuer C.F. von Weizsaecker.* Goettingen: Vandenhoeck & Ruprecht, pp. 98-118.

Sneed, J.D. (1971, 2nd ed. 1979). *The Logical Structure of Mathematical Physics.* Dordrecht: Reidel.

Poznań Studies in the Philosophy of the Sciences and the Humanities
2000, Vol. 72, pp. 173-197

C. Ulises Moulines

IS THERE GENUINELY SCIENTIFIC PROGRESS?

ABSTRACT. The traditional notion of scientific progress is too simple-minded and has justifiably been questioned by epistemic relativists. Genuinely scientific progress cannot be explicated as "knowing more about the same things". History and semantic analysis speak against such an interpretation. However, the epistemic relativist is wrong in inferring from this that the notion of genuinely scientific progress should be abandonned. Using the structuralist metatheory allows for several explications of scientific progress which are both conceptually precise and historically applicable. The core idea is that we may have theoretical incommensurability in the course of scientific development and nevertheless a specifiable structure of comparability at the level of intended applications.

0. The problem

The question posed in the title of this essay has been worrying philosophers of science since the publication of Thomas Kuhn's *Structure of Scientific Revolutions*. It is a vexed question for the philosophy of science because, in spite of its seemingly elementary character and contrary to appearances, it does not allow for a straightforward, simple answer.

In view of the overwhelming historical data about the evolution of science since Antiquity, a negative answer to the question posed here appears, at first glance, as the frivolous *boutade* of a snob, so that our common sense very strongly feels inclined to exclaim: "But of course there has been a tremendous progress in scientific knowledge!" However, if we go on asking what good reasons we might have for this positive answer, we have to undertake a careful analysis of what *genuinely* scientific progress really amounts to. And then, through such an analysis, we might be led, in turn, to wonder whether the frivolous snob is not right after all. At least, we may then feel that we are not entitled anymore to take the positive answer for granted.

1. What is the general structure of progress

To clear up the issue, let's start with some truisms. In spite of their trivial character, they have to be considered explicitly because the discussion about scientific progress has become so muddled.

First, to ask whether there has been genuinely scientific progress is *not* to ask whether there has been some sort of *moral* progress. For some reason or other, we might come to the conclusion that science, as we know it, is a bad thing. So, if there is scientific progress, then there is moral regress. More generally, even if we were convinced that there is genuinely scientific progress, this does not imply anything about "progress of human civilization" (unless we equate science with civilization – something I would not stand for). Our question is whether there is scientific progress in a clearly specifiable sense – not whether this is good or bad for mankind.

Another trivial remark I want to make is this. When we ask questions about progress in any given field, we have got to compare certain items in this field in a diachronic perspective according to some ordering criteria. That is, we ask ourselves whether we can put those items in a certain ordering relation – from less to more in historical time. It is not necessary for the construction or detection of this ordering that *only one* criterion for "more than" or "better than" is used. There may be several relevant criteria and they may even stand in a not quite harmonious correlation. But this doesn't matter much as long as we can show that they all somehow combine in a way that allows us to state that there is an *overall* trend of "bettering". For example, I start learning a foreign language and, in the course of time, I ask myself whether I have made some progress. To answer this question, I might consider whether I know more words than before, or whether I can understand better what native speakers say, or whether I can construct sentences in a more correct way, and so on. These criteria are not necessarily correlated in a simple way. For example, I might find out that I have learnt many more words but am equally bad as before at mastering the rules of grammar. Genuine progress must be conceived as an overall trend according to all relevant criteria. If I claim that I have made tremendous progress in learning that foreign language because I know ten times more words than before, but still don't know how to put them together to form a single correct sentence, then nobody will take my claim seriously. This would not be *genuine* progress.

Third remark. Progress in a given field may be enhanced by some capacities in another field. But this does not imply that, if we have progress in the first field, we also have progress in the second. I might make good progress in learning a foreign language thanks to my good memory. But my memory doesn't necessarily get better as I learn the language. It may be always the same, or even get worse, or at any rate it may be irrelevant whether it gets better or not.

Science is a cultural process. When we ask about genuinely scientific progress we ask whether we can detect an overall ordering relation in this process which comes out from applying several significant criteria in combination. The answer will therefore depend both on empirical historical data about the different stages in the development of science since Antiquity, and on a conceptual analysis of how the relevant ordering relation is to be conceived. This is the reason why the issue of scientific progress depends both on the history and the philosophy of science. The present essay deals with the issue mainly from the latter perspective.

There are a number of cultural processes where genuine progress, cannot be seriously doubted, where the relativist denying progress in the field immediately reveals him/herself as what s/he is: a frivolous snob. One example is technology. If we compare the devices invented by people for transportation in diachronic perspective, and consider speed and ease as criteria for the ordering relation, then there is no doubt genuine progress from the Neolithic to the era of the Jumbo jet. The relativistic snob clearly is not a serious discussion partner here. Another example is medicine, or rather: medical therapy. To deny that penicillin is more effective than Galen's therapy for healing and/or postponing death, is completely uninteresting even as a joke. Demographic statistics speak for themselves. (Again, this doesn't imply that therapeutical progress is a "good" thing; it may be argued that it is a *regressive* trend that so many people live so long; however, we are not discussing "progress of mankind" in general but progress in a specific domain.)

So, there are specific areas of human culture where genuine progress cannot be denied seriously. A clear notion of progress together with the relevant empirical data makes this claim as well-founded as any empirical matter of fact can be.

On the other hand, it is equally clear that there are domains of human culture where the assertion of genuine progress would be highly problematic, to say the least. One example is religion. There has certainly been a succession of religious systems in different parts of the world at different times. But the question here is not just whether there has been genuine progress but rather: Does it make sense to ask for progress here? From the 7th century on, a great portion of what Braudel has called "the Mediterranean civilization" changed from Christianity to Islam. Was this progress? For a Muslim, the answer is clearly positive; for a Christian, the process was a terrible backlash; while the "freethinker" may look at the whole issue with a condescending smile... The "objective" answer to the question "Why do you believe that Islam is better than Christianity?" can only be: "Because I *am* a Muslim". There are no independent criteria by which we can assess the passage from Christianity to Islam as a "bettering" or a "worsening". We have here the kind of circular arguing which makes the application of the idea of progress nonsensical.

Another example is art. We can certainly compare different works, within the same school and come to the conclusion that there was progress – with respect to the standards set by *that* school. But what does it mean to say that there was genuine progress in the transition from representational to abstract painting? In what sense are Kandinsky's best works *better* than Manet's best paintings? The overall comparative ordering we need for the notion of progress doesn't seem to appear here.

Therefore, there also are important kinds of cultural evolutions where the relativist is right and the notion of progress scarcely makes any sense. Now the question about scientific progress is, to put it in a nutshell, whether scientific development belongs to the first or rather to the second group of cultural developments. Is science more like technology and therapy, or is it rather like art and religion?

Before Kuhn and Feyerabend came on the scene and, in the sequel, were followed by some neo-Wittgensteineans, social constructivists, ethnomethodologists, the "Edinburgh school", Rortyans, francophile postmodernists, and the like – let's call all of them "epistemic relativists", the answer to the previous question seemed to be quite straightforward: science is like technology and therapy, and not at all like art or religion. Still more: science, technology, and therapy constitute a coherent unit – "the modern scientific mind" – and only with respect to this unit can we speak of genuine (epistemic) progress. This is still the common opinion of the (semi)cultivated man-on-the-street everywhere in the world (and not only in Euro-American countries), and it is also the common opinion of the vast majority of practising scientists, who never heard about Kuhn or don't care about what he said. For all these people, it is quite clear that the development of science has resulted in an increase of objective knowledge, and this, in turn, has produced an increase in technological power and therapeutic capacities. Let's call the (explicit or implicit) defenders of this view "epistemic progressives". Here we have the two main camps of contemporary *Weltanschauungen*: On the one hand, the epistemic relativists, represented mainly in sophisticated circles of history, philosophy and, above all, sociology of science; on the other hand, the epistemic progressives, represented by a few philosophers, by a great number of natural scientists and, above all, by "cultivated common sense". The first group questions the idea of scientific progress; the second just ignores this questioning or, at best, views it as just one more of these romantic-obscurantist fashions which, for the last two hundred years, have periodically annoyed civilized mankind but have ultimately vanished into oblivion. Who is right? Or: Can we discuss the issue by means of arguments or is it just a matter of faith?

2. Some (bad) arguments for epistemic progressivism

Both relativists and progressives have offered some arguments for their views. Some of these arguments are better than others. Since I rather belong to "the progressive camp" (i.e. I think one can introduce a meaningful notion of genuinely scientific progress and produce historical examples where this notion applies), I will concentrate on what I consider to be the most serious challenge to epistemic progressivism coming from the relativistic camp: the incommensurability thesis. I will try to show that there is a sound core in the incommensurability thesis and that this sound core is enough to block some simple-minded progressive arguments. But then I will also show that one can construe a plausible notion of scientific progress immune to incommensurability and probably instantiated in a great number of cases.

Before I come to deal with incommensurability, however, let me say a few words about three epistemic-progressive arguments which are found in the literature and which I consider rather poor.

The first argument for epistemic progressivism may be labelled "the technological argument". It is probably the first argument coming to the mind of the cultivated common sense. It runs like this: technology (I include here therapeutic "technology") is inseparable from science; there is undoubtedly technological progress (as we have already seen); therefore, there is scientific progress.

The argument has two flaws: an empirically false premise and a conceptual *non-sequitur*. It is just false that technology is inseparable from science. There are plenty of historical examples, not only in ancient history but also in modern times, of technological developments which were quite independent of contemporary or previous scientific developments. (It would be ridiculous to give a list of examples but just think of the well-known slogan "science owes more to the steam-engine than the steam-engine to science".) On the other hand, there are plenty of scientific "paradigm changes" which were independent of technological innovations. (Just think of the inception of general relativity theory: What technological invention prompted it?) So much for the historically false premise. As for the *non-sequitur*, it is this. Even if it were the case that a great deal of technological developments rested on science, it still doesn't follow from the fact that we can detect genuine progress in the first case that there must also be genuine progress in the second. The example of language acquisition illustrates this. Even if I need memory to make progress in learning a foreign language, I need no progress in my memory capacity. It may be the case that some technological advances need science (and even scientific *change*), but why should they need specifically scientific *progress*?

All this is not to deny that we may discover significant correlations between scientific and technological progress as soon as we have introduced a plausible, but independent, notion of scientific progress. I will come back to this point later on. At any rate, there is no well--founded way to extrapolate from the assured fact of technological progress to the hypothetical idea of a genuinely scientific progress.

Another argument for epistemic progressivism is a characteristic of those philosophers of science usually called "scientific realists". It rests on the idea of explanatory success. It runs like this. The main aim of (theoretical) science, with or without technology, is to explain phenomena; there has been a steady increase in the *explanatory success* of scientific theories since ancient times; *ergo*, there is scientific progress.

Now, the foregoing argument is a clear case of a *petitio principii*: it just presupposes that what counts as a "good explanation" in a new scientific theory *T'* would also have counted as a good explanation in a previous scientific theory *T*. Kuhn, Feyerabend, and the others rightly pointed out that precisely this presupposition usually doesn't hold. Aristotle would have been appalled by the way Newton explained the falling of the apple from the tree – with all its brutal idealizations, contrary-to-fact assumptions, mathematical tricks, and so on. We can compare scientific theories with respect to their explanatory successes only when they belong to the same scientific tradition (the same "disciplinary matrix", as Kuhn puts it, or the same "research program" in Lakatosian terms). But this is like comparing the relative merits of different paintings within the same artistic school: such a comparison makes sense but no general idea of progress is derivable from it.

A different but related argument some scientific realists use for their belief in progress is the *convergence* thesis: the development of science is such that scientific theories, in spite of temporary difficulties and drawbacks, *in the long run* converge to a final, ideal theory which is supposed to be completely successful. This convergence *is* scientific progress.

The problems with this scientific eschatology are similar to those a non-believer has when confronted with a literal interpretation of Christian eschatology: First, there is no rational ground for assuming the coming of such a Last Judgement – history of science, as much as the history of mankind at large, appears to be a rather chaotic, purely contingent affair. Second, even if we could make sense of the idea of an endpoint in scientific development, there is no warrant for the assumption that this must be a "happy end" in the sense of an intuitively plausible notion of progress. The final theory to which all others converge may just be worse according to previously established and intuitively plausible criteria for the relation "better-than" between scientific theories. The feats of the human spirit might converge... but converge to *hell*.

3. What is right about incommensurability

The arguments just examined are among the most popular for epistemic progressivism. However, they fall short of being convincing. Of course, this still doesn't imply that epistemic progressivism is wrong. But their failure may already cast a shadow on our naive belief in genuine scientific progress. The shadow becomes much larger when we get confronted with the thesis of incommensurability. The epistemic progressive has to take this challenge seriously.

It is important to note right from the beginning that, even if the incommensurability thesis were true, this wouldn't imply that there is no scientific progress. What the incommensurability thesis questions is a *specific* understanding of scientific progress: the idea that scientific progress consists in *knowing more and more about the same*. But there may be other plausible notions of scientific progress which are immune to the phenomenon of incommensurability. As a matter of fact, this is one central claim of this essay: that in a plausible sense (still to be specified) there is *both* incommensurability *and* progress in science.

Since Kuhn and Feyerabend (independently) started arguing for incommensurability, thousands of pages have been written on it – pro and con. I assume the reader is well-acquainted with the essentials of this discussion. The standpoint I take towards this issue has been laid out in (Balzer *et al.* 1987) and in somewhat different terms in (Moulines 1991). Here, I intend to filter out what I take to be the sound core of the incommensurability thesis when we free it of superfluous ballast.

Pointing to many case studies in the history of science, Kuhn and Feyerabend emphasized a fact that had already been noticed by previous scholars (Fleck, Koyré and Bachelard, particularly): the development of science, even within a single, well-established discipline, is not at all "smooth". There are more or less deep *cuts* in it, and the transition from the stage before the cut to the stage after it doesn't seem to be determined essentially by rational argument and cool experimentation. These cuts ("revolutions" or "paradigm-shifts" are the usual labels for them) are accompanied by curious socio-psychological phenomena: sudden conversions, the use of power instead of rational discussion, propaganda instead of arguments, highly emotional controversies, and the like. Clearly, these are things a genuine epistemic progressive must profoundly dislike. Indeed, during such phases, scientific change looks more like the change from Christianity to Islam, or from representational to abstract painting, than like the transition from propeller airplanes to jets. Kuhn and Feyerabend said that theories separated by such cuts are "incommensurable", in quite the same sense as religious systems or artistic styles can be said to be incommensurable. And this is exactly what the epistemic relativist had been longing for and the epistemic progressive had been afraid of all the time.

Now, the thesis of incommensurability has often been understood as merely the assertion that there are such cuts and that they are accompanied by those disturbing socio-psychological phenomena. And epistemic progressives have then set out to "prove" that the historical examples put forward by incommensurabilists are not *that* dramatic, that some elements of rational theorizing and experimenting *did* play a role in even the most revolutionary changes – in a word, that the difference between peaceful scientific development and dramatic changes of paradigm is just a matter of degree. But this amounts to trivializing the issue at stake. It may well be that the difference between the most conservative scientific development and the most dramatic change of theory is just *a matter of degree*; however, such a gradualism does not attack, just conceals, the interesting questions about the nature of scientific change. If you compare life in Sarajevo with life in Andorra in 1993, you can also say that the difference is just a matter of degree, because you can devise a long chain of intermediate instances leading from one to the other, each pair of similar instances differing only "a little"; but building such a gradual chain from Sarajevo to Andorra wouldn't clarify the crucial issue. It wouldn't help you a lot when trying to understand what was really happening in Sarajevo and what the nature of war is. Gradualism about scientific change may be the right view but it really doesn't answer the challenge posed by incommensurabilists. Incommensurabilities between scientific, as well as political, systems may be gradual but this fact doesn't explain them away nor does it help us to understand the nature of the conceptual structures giving rise to them.

On the other hand, just taking notice of the phenomenon of drastic scientific breaks is not enough. Merely asserting that there are cuts in scientific development – some deeper than others – and that these cuts go hand in hand with apparent breakdowns in scientific communication, is not the really interesting aspect of the thesis of incommensurability – at least from an epistemological point of view. We should instead take this thesis as a semantic, metascientific theory; by a sort of "abductive argument", it purports to *explain* those phenomena labelled "scientific revolutions" – be they gradual or not. That is, *if* we assume the thesis of incommensurability as a fundamentally semantic-epistemological theory about science in diachronic perspective *then* we can explain much better the undeniable existence of periodic, more or less profound scientific crises and the concomitant socio-psychological phenomena. In particular, we can explain why, during such periods, scientists "don't understand each other", why what we usually conceive as the capacity for "rational argumentation" becomes considerably diminished.

The metatheoretical explanatory core of the incommensurability thesis is what Feyerabend and others called "radical meaning variance": after a scientific revolution, the meaning of the central terms of the theories involved change so drastically that there is no guarantee anymore that people talk about the same, and if this is the case, then there is no guarantee either that people have come to know *more about the same things* – which is the mark of scientific progress according to the naive epistemic progressive. On the contrary, after such a change, there is every reason to think that the new scientific paradigm does *not* deal with the same objects as the previous one – it has constructed a different "universe of discourse" for its central terms; and the second universe of discourse doesn't include the first one. In such a case, it is plain that there has been *no* progress in the sense of knowing more about the same.

Note that in order to draw this conclusion it is not necessary to have a highly sophisticated theory of meaning for scientific terms. All we need is the following mini-theory, which seems to be the common property of a great deal of current theories of meaning. It says that the meaning of a scientific term is constituted by at least two "vectors":

a referential and a non-referential component. The non-referential component (be it Frege's sense, Carnap's intension, Putnam's stereotype, or whatever) may itself be a quite complex unity, depending on other semantic categories. We only need to assume that the non-referential component is relevant for the epistemic access to the referent. (According to Frege, the sense is *the way* by which we become acquainted with the referent.) Let's call this semantic mini-theory "the theory of the epistemic relevance of the non-referential for the referential component of meaning" ("ERNR"-theory, for short)[1].

Now, consider any central term of a scientific theory. The non-referential component of its meaning – leaving aside any presystematic, common-sense associations – is essentially given by the theoretical context in which the term appears. This context, in turn, is basically constituted by the lawlike relationships connecting the term in question with other central terms of the theory. A more or less drastic scientific change is precisely characterized by a more or less drastic change in those lawlike relationships, and therefore in the possibilities of epistemic access to the "real referent". The referent of the term which has gone through a paradigm-shift *may* be the same[2] but since our way of getting acquainted with it has radically changed, we cannot know whether we now know more or less about it than before. We cannot "weigh" the different ways of epistemic access to the referent before and after the paradigm-shift because we are lacking, *ex hypothesi*, the *tertium comparationis*. In short, we cannot *know* whether we know more about the same thing. There may be scientific progress but we will never be able to find out whether there has actually been any – which is a rather strange thing to say about scientific progress. This conclusion, of course, just makes apparent that, assuming ERNR and the historical existence of more or less drastic theory changes, the idea

---

1 For reasons laid out in (Moulines 1991b) I don't believe in the causal theory of reference. But even an orthodox causal theorist of reference, it seems to me, could accept ERNR. Indeed, we may grant him/her that the referent of the English term 'water' today is the same as the referent of its philological ancestor in some remote Indo-European language, where the term was allegedly introduced for the first time through an "introductory event" or "naming ceremony" (Putnam), and that this referent has survived all changes in our beliefs about water. Nevertheless, if we are to have epidemic access to the object 'water' "really refers to", that is, if we are to **know** what water really is, then this must be by means of the non-referential component of 'water', e.g. through the sense of 'water' conveyed by a chemical theory.

2 If the causal theorist is right; if he is not, we are not entitled to even assume that much.

of scientific progress as knowing more about the same vanishes into thin air. This, I think, is the sound core of the incommensurability thesis.

4. Comparable incommensurabilities

From our rather harmless, semantic-epistemological mini-theory together with some almost indubitable historical evidence we may reasonably conclude that there is no scientific progress in the sense of knowing ever more about the same; or, at best, we have to conclude that this notion refers to a mysterious kind of Kantian unknowable "*Ding-an-sich*". The idea of knowing ever more about the same is an uncontrollable speculation at best, a nonsensical assertion at worst. But this still is no reason for the epistemic relativist to triumph. We may have both incommensurability *and* an ordering of theories in diachronic perspective which intuitively corresponds to the general notion of progress we have sketched at the beginning of this essay. As a matter of fact, the two original proponents of the incommensurability thesis, Kuhn and Feyerabend, never denied the possibility of having a sensible notion of scientific progress – a fact that should disturb the later relativists they inspired. The reason for this is that both Kuhn and Feyerabend admitted the possibility of comparing incommensurable theories, and if you can compare theories according to some epistemologically relevant criterion, then you may also be able to say which is the *better one* and build up the ordering you need for the idea of a genuinely scientific progress. That incommensurability doesn't imply incomparability was already emphasized by Kuhn and Feyerabend a long time ago. In 1976, Kuhn wrote:

> Most readers of my text have supposed that when I spoke of theories as incommensurable, I meant that they could not be compared. But 'incommensurability' is a term borrowed from mathematics, and it there has no such implication. The hypotenuse of an isosceles right triangle is incommensurable with its side, but the two can be compared to any required degree of precision. What is lacking is not comparability but a unit of length in terms of which both can be measured directly and exactly[3].

And, in spite of Feyerabend's efforts to disentangle his own incommensurability concept from Kuhn's, we could read a similar declaration in his review article of 1977 on Stegmüller:

---

[3] Cf. (Kuhn 1976, p. 191).

> When using the term 'incommensurability' . . . I never inferred incomparability from it. . . . Quite the contrary, I tried to find means of comparing such theories. Comparison by *content*, or *verisimilitude* was of course out. But there certainly remained other methods[4]

Curiously enough, these passages haven't been paid much attention to either by critics or sympathizers of the incommensurability thesis. One reason may be that Kuhn and Feyerabend themselves were not quite explicit about how the comparison of incommensurable theories should proceed or about how they proposed to explicate the relation scheme "T is incommensurable and comparable with T'". Kuhn's analogy from geometry is not very illuminating: in the case of scientific development, we have nothing like the system of real numbers which may be used as a common base to compare two incommensurable items and tell which is "longer". As for Feyerabend, he just states the existence of some "methods" of comparison but doesn't tell us much about them. We are still confronted with the problem of what it really means for two theories to be incommensurable *and* comparable.

The uncompromising epistemic relativist need not be quite impressed by the possibility of having incommensurable and comparable theories. Not every comparability is relevant for the idea of progress. He/she might exclaim: "Of course pairs of theories separated by a historical cut are comparable in some sense. For example, you may compare their respective lifetimes and come to the conclusion that one *lasted much longer* than the other. Or you may compare them with respect to geographic areas and see that one of them was popular in *more countries* than the other. So what? What claim about scientific progress can you derive from that?"

This challenge of the epistemic relativist too has to be taken seriously. So long as we don't set up criteria of comparison between succeeding theories which clearly are *epistemologically relevant*, the claim that we can have incommensurability and comparability will be devoid of significance for the question of scientific progress. The problem is far from trivial and I don't pretend to have the definitive solution. However, in the last part of this essay some relevant criteria of comparison will be advanced which can contribute to make the idea of genuinely scientific progress more plausible.

---

[4] Cf. (Feyerabend 1977, p. 365).

5. Scientific progress without common reference

There is a sound core in the incommensurability thesis. But there is a sound core in the naive belief in scientific progress as well. Even if we leave aside science-based technological progress there still is "pure" *scientific* progress. The relativistic denial of it is *really* a snobbish denial after all. It cannot be denied seriously that *there* is an ordering relation through historical time which intuitively corresponds to the relation "better-than", at least in some disciplines. In astronomy, we can detect such an ordering in the line Ptolemy-Copernicus-Kepler-Newton-Einstein – even if the main technological application (navigation) of this theoretical sequence was not particularly improved by the development. The serious problem is how to explicate this sound intuition. Assuming a transtheoretical, fixed set of referents through historical time about which we know more and more will not do. Incommensurability is there to block this way out.

To find an alternative way out we have got to be more precise now about the nature of the units to be ordered and compared in scientific development. If we leave aside technological improvement, these units can only be theories. In principle, another possibility for detecting genuinely scientific progress could be the improvements in experimental techniques[5]. However, changes in experimental techniques may be characterized as actual improvements only with respect to the given theories which *make sense* of them. Experimental techniques are subsidiary to theories. Otherwise, changes in experimental techniques would just be like substituting one game for another. For some reason or other, we may get bored playing backgammon and start playing chess. Even if one game requires more skills than the other – why should we speak of genuine progress here?

So, the *basic* units to be compared when speaking of genuinely scientific progress have got to be theories. The question is: how can we define a plausible notion of "better-than" between theories which on the one hand doesn't amount to the fact the one theory is used to devise better machines than the other, and which, on the other hand, don't depend on the assumption that both refer to the same kind of "things" and one allows us to know more about them than the other?

To answer this question, we need to be more precise, in turn, about how to conceive scientific theories. Until recently, it was quite common among philosophers of science to conceive theories as sets of statements. After the "semantic turn" in philosophy of science, this

---

[5] (Hacking 1983) may be interpreted as supporting this view of scientific progress.

view is not so popular anymore, but it still is surreptitiously quite influential.

There are many reasons speaking against the statement view of theories but one of them is certainly that it doesn't allow for an adequate treatment of the issue of scientific progress. The reason is quite simple. If theory *T* is a set of statements (ideally: an axiom system) and theory *T'* is another set of statements (another axiom system), then it makes sense (epistemologically) to compare *T* with *T'* and see whether *T'* is better than *T* *only if* it is guaranteed that *T'* contains more true statements, or more truth-like statements, *about the same things* than *T*. Otherwise, even if we came to the conclusion (which in real-life cases we won't be able to reach, anyway) that *T'* contains more true, or truth-like, statements than *T*, this would not warrant any conclusion about a genuinely scientific progress. A new sports magazine may contain more true statements than *The Wall Street Journal* but in what sense is it *better*? Yet the assumption that the statements contained in *T'* speak about the same things as the statements in *T*, is precisely the assumption we aren't entitled to make if the results of the foregoing discussion are accepted.

Consequently, we should abandon the statement view of theories, especially when we deal with the issue of scientific progress. Trying to tell which theory is better by comparing their respective statements leads to a blind alley[6]. Under the circumstances, it doesn't seem too far-fetched to tackle the issue from the point of view of a decidedly non-propositional conception of theories, and this is the particular version of the semantic approach widely known as "the structuralist view of theories", or "structuralism", for short. In particular, we may explore the possibility that this conception of theories leads us out of the dilemma described above (scientific progress is either just technological progress or else knowing ever more about the same).

Here, I must presuppose the reader is acquainted with the essentials of structuralism[7]. The structuralist methodology is rather technical;

---

6 This was already seen by Stegmüller in his *Structure and Dynamics of Theories* some twenty years ago. But his line of argument was somewhat different, and not much attention was paid to it anyway; so this insight has to be renewed time and again.

7 For more details, see the "standard" work on structuralism: (Balzer *et al.* 1987). There is also a detailed "Bibliography" of contributions of structuralism to philosophy of science, both in general and in particular, up to 1994 (Diederich *et al.* 1994).

however, to make the first steps towards a clarification of the present issue I will restrict the technicalities to a bare minimum.

According to structuralism, the cultural products usually known as "scientific theories" are to be conceived not as a set of statements but rather as complex (non-propositional) structures. Roughly, a scientific theory is to be conceived as a *hierarchical net* – more concretely: a tree-like net. The nodes in the tree are structural units known as "theory-elements" and, the branches connecting the nodes are determined by a relation of specialization. Usually, there is a "top" element in the tree (the "fundamental law(s)" of the theory) out of which everything else (the "special laws") is constructed by specialization. Each single theory-element is an ordered pair of two entities: a (formal) core *K* and a domain of intended applications *I*. *K*, in turn, contains a number of different components, all of which can be defined in terms of (formal) model theory; on the other hand, *I* can only partially be described model theoretically since its determination *also* depends on highly pragmatic notions. In particular, it depends on the actual *interests* of the theory-user (normally: a collectivity). In many cases, the interests determining *I* are directly or indirectly technological; but this need not be always the case. In many examples of actual scientific theories, *I* will be determined by technological as well as non-technological interests, and in some other examples there will be no technological interests at all. The only important thing to notice about *I* here is that the pragmatic aspect of its determination is given by the "vital" interests (whatever they may be) of the collectivity of theory-users. Indeed, *I* and the collectivity constitute an indissoluble unity. We may say that *I* is what gives sense to the "collective form of life" of the corresponding theory-users, their *Lebenswelt*. (A neo-Wittgensteinean or a constructivist may rejoice here.) The collectivity invents *K* precisely because it wants to come to grips with *I*.

The determination of *I* depends on pragmatic, non-formalizable notions like "vital interests of collectivity *C*" (where *C* is a historically changing group of individuals of the species *Homo sapiens*). But this is only one aspect of the determination of *I*. Not all aspects are opaque to formal treatment. In order "to come to grips" with *I*, *C* conceives it in terms of (some of) the central concepts of *K*: *C K*-conceptualizes *I*. In this endeavour, *C* may be more or less successful.

It is very doubtful that this notion of the intended applications of a theory, which is partly pragmatic, partly model-theoretical in kind, can be (much less: *should* be) explicated in terms of a stable "world" of fixed objects which are the referents of a universally understood language. At any rate, nothing in the following analysis presupposes such a notion.

Starting with an extremely simplified picture of what *K*'s components according to structuralism are, we may assume that *K* just consists of two units: a class of "potential models", M_p, and a class of "actual models", *M*. (Both are definable by means of set-theoretical predicates.) When we say that *I* becomes *K*-conceptualized, we mean that *I* is determined in such a way that the intended applications of the theory are seen as *potential* models of it. The "borders" of *I* may depend on strongly pragmatic factors, but *C* must assume, at any rate, that $I \subseteq M_p$. Under this assumption, if *C* finds out that $I \subseteq M$ as well, we say that the theory has been completely successful. (In real-life cases we will get less than this: we will have to be satisfied with $I \cap M \neq \emptyset$ so that $I \cap M$ "is considerably bigger" than $I - M$. A theory where this happens will normally be considered a "good" theory – whatever Popperians may object.)

A metatheoretical hypothesis put forward by structuralism is that we will be able to distinguish, in a plausible way, two different conceptual layers in any "normal" empirical theory: the layer of the *T*-theoretical and the layer of the *T*-non-theoretical concepts. The first can only be determined within the theory, the second comes "from outside", normally from *underlying theories*. For example, in mechanical theories, dynamical concepts are typically *T*-theoretical, whereas kinematical concepts are *T*-non-theoretical. Let's assume this metatheoretical hypothesis is right – at least for most interesting theories where the issue of progress makes sense. Then, the former story about the relationship between *I* and *K* becomes a bit more complicated. We have then to define the class M_{pp} which is the model-theoretical reduct of M_p to its *T*-non-theoretical substructures[8]; and then we should conceive *I* as $I \subseteq M_{pp}$ only. Within this framework, the theory is completely successful whenever $I \subseteq r(M)$. (Again, in real-life cases we must content ourselves with finding out that $I \cap r(M)$ is "considerably bigger" than $I - r(M)$.)

Up to this point, we have laid out what it means for a theory to be successful in a particular theory-element $T = \langle K, I \rangle$. But we also have said that, normally, a theory will consist of several theory-elements ordered in a net. Each theory-element in the net has its own domain of intended applications. Nevertheless, we can introduce also the no-

---

[8] Formally, this means that we construct (according to our *T*-theoreticity criterion) a many-one function $r: M_p \rightarrow M_{pp}$. To each full structure of *T*, *r* applies its corresponding *T*-non-theoretical substructure.

tion of a "global" domain of intended applications for the whole net. The question of how to define this global domain in a way which is both technically correct and intuitively plausible, is far from being trivial. But it can be solved in a satisfactory way too[9]. Therefore, we have the notion of an I_T for a single theory-element as well as the notion of an I_N for a whole net, i.e. a whole theory.

Let's see how the previous notions can be switched into the issue of scientific progress. We make just two elementary assumptions: 1) theories usually are tree-like theory-nets with a top element; 2) theories usually have two conceptual layers (the theory's specific one and the one coming from "outside"), symbolized by M_p and M_{pp}, respectively. Under these assumptions, different kinds of scientific progress can be distinguished – at least *three*. They can be explicated in quite precise terms within the framework of structuralism. Let's consider them.

6. Lakatosian progress

The first obvious kind of genuinely scientific progress may be called, in honour of Imre Lakatos, "Lakatosian progress": it is progress within a research tradition. In this case, the two theories to be compared are two different theory-nets sharing the same "hard core", i.e. – in our terms – the same core on the top element (call it K_o) and the same global I_N. Call the first theory N and the second N'. Suppose N' succeeds N in time. Then, we will say that there has been genuinely scientific progress in the transition from N to N' whenever, roughly speaking, two conditions hold – one on the T-theoretical and one on the T-non-theoretical level: first, the laws in N' must be at least as informative as the laws in N; second, C must be at least as successful when applying the laws to I_N in N' as it is when doing the same in N.

Let's try to make this formulation more precise by using elementary set-theoretical means. First, let's "fragment" the global domain I_N into the subsets $I_1, I_2, \ldots, I_n$ in N, corresponding to the different theory-elements ("special laws") $T_1, T_2, \ldots, T_n$ of which N consists. (Idealizing a bit, we may assume that $I_1 \cup I_2 \cup \ldots \cup I_n = I_N$ – though this is not essential to the argument.) Second, use the symbol 'σ' to express the

[9] This question is discussed at length in (Balzer *et al.* 1987, Chs. IV and VIII). A more sophisticated and elegant solution to this problem has been provided by U. Gähde in a still unpublished manuscript.

relation of specialization between theory-elements[10]. Further, define the sets $S_i =: I_i \cap r(M_i)$ for $1 \leq i \leq n$ ('S' stands for 'success'). Finally, define the corresponding sets S_i' *mutatis mutandis* with respect to the theory-elements $T_1', T_2', ..., T_m'$ in N'. Then, we will have a clear case of *Lakatosian progress* whenever following conditions hold:

$$(L) \quad (a)\ \forall T_i'\ \exists T_j\ (T_i' \sigma T_j) \qquad 1 \leq i \leq m \quad 1 \leq j \leq n$$

$$(b)\ \forall S_i\ \exists S_j'\ (S_i \subseteq S_j') \qquad 1 \leq i \leq n \quad 1 \leq j \leq m.$$

Intuitively, condition (L)-(a) says that special laws are more specialized (more "accurate") in N' than in N; (L)-(b) says that N' is more successful. In real-life cases, the set-theoretical relationships won't usually be so neat: we may have to replace the strict universal quantifier in (b) by an "almost all" operator or something still more complicated but I don't want to get involved in these technicalities here; I hope the main line of the argument has been made clear through the foregoing ideal conditions[11].

For this first kind of genuinely scientific progress we have assumed that the conceptual structure of the succeeding theories, at least at the top level K_0, stay being the same. The epistemic relativist will of course object here that this is the kind of scientific change he/she is not interested in. S/he may concede that there may be a number of historical examples that can be reconstructed in this way[12] but that these are just cases of "normal science" – not of paradigm-shifts. All right, let's see, whether a second kind of genuinely scientific progress fits the intuitions of the relativist better.

---

[10] This relation is formally defined in standard expositions of the structuralist metatheory, e.g. in (Balzer *et al.* 1987, Ch. IV.1).

[11] More technicalities should be introduced into the foregoing formulation if the full apparatus of structuralism is employed and, in particular, if we take approximations into account – by no means a trivial matter. But the reader well-acquainted with the full-grown theory-concept laid out in (Balzer *et al.* 1987) may fill in the details missing here in a straightforward way.

[12] Those given by Lakatos himself, for example – other examples of a similar sort are to be found in (Balzer *et al.* 1987, Ch. V).

7. Progress with partial incommensurability

Another possible situation is this. N and N' share a domain of intended applications, or at least their respective domains can be described in commensurable terms, but the respective theoretical layers are completely different; that is, N and N' share M_{pp} but differ with respect to M_p. In such a case, it is obvious that we can't say that they share at least K_0. Here we have a kind of *partial incommensurability*: incommensurability on the T-theoretical level and commensurability on the T-non-theoretical level. However, in this case we may still speak of progress in a quite precise, straightforward sense. Actually, it is essentially the same kind of progress as above when we leave the T-theoretical layer out of consideration. Since M_{pp} is common to both theories, we may define the sets S_i and S_j' in the same way as above and use the same condition (L)-(b) to define progress. We may call this "progress with simple partial incommensurability" (SPI):

(SPI) $\quad (a) \equiv (L)$-(b).

There is some incommensurability here because (L)-(a) is missing but there also is some methodologically relevant comparability – the one expressed in (SPI)-(a).

In order to simplify the analysis of partial incommensurability it has been assumed so far that here we have $I_N = I_{N'}$. This, of course, is a somewhat idealized assumption. Strictly speaking, it would not hold in many cases where, intuitively, we would like to use the notion of partial incommensurability. Let's examine more complex (and more realistic) possibilities.

In the first place, even though I_N and $I_{N'}$ might be describable by means of the terms of the same M_{pp}, it is not necessary, for a more flexible notion of scientific progress, for us to have a strict identity between I_N and $I_{N'}$. We may have $I_N \neq I_{N'}$, but $I_N \cap I_{N'} \neq \emptyset$. Take $I^* =: I_N \cap I_{N'}$. Then we can see whether a) above holds for I^* and define progress accordingly. Furthermore, consider $\bar{I}_N =: I_N - I^*$ and $\bar{I}_{N'} =: I_{N'} - I^*$. If we find out that $\bar{I}_N \cap r(M_0)$ is "smaller" than $\bar{I}_{N'} \cap r(M_0')$ and $\bar{I}_N - r(M_0)$ "bigger" than $\bar{I}_{N'} - r(M_0')$ then we may say that we have a clear case of scientific progress through partial incommensurability. Let's summarize these conditions in the following definition of "progress with partial incommensurability" (PI):

(PI)

(a) $\exists I^* \ (\varnothing \neq I^* = I_N \cap I_{N'})$

(b) $\forall S_i^* \ \exists S_j'^* \ (S_i^* \subseteq S_j'^*)$
$\left(\text{where } S_i^* =: I^* \cap I_i \cap r(M_i) \text{ and } S_j'^* =: I^* \cap I_j' \cap r(M_j')\right)$
$1 \leq i \leq n \quad 1 \leq j \leq m$

(c) $\left\| \bar{I}_{N'} \cap r(M_0') \right\| > \left\| \bar{I}_N \cap r(M_0) \right\|$

(d) $\left\| \bar{I}_N - r(M_0) \right\| > \left\| \bar{I}_{N'} - r(M_0') \right\|$[13].

The essential point in the idea of progress through (simple) partial incommensurability is that, even though the respective domains of applications are not exactly the same, they may be "inserted" into the same conceptual framework (M_{pp}). It is this framework which allows for a methodologically relevant comparison.

Even this last point can be handled in a more liberal way without breaking the main stream of the argument. The statement that M_{pp} keeps being the same entity across scientific change should not be taken too literally. As a matter of fact, the *linguistic setting for* M_{pp} may change more or less drastically, but this is not essential for the kind of scientific development we are considering. Structuralism is *not* a linguistic approach to science anyway: one of its central claims is precisely that, when reconstructing scientific theories, the actual *terms* used are not very important as long as you can make clear that the *structures* determined by those terms are essentially the same. "Essentially the same" means here that you can make explicit a set-theoretical correspondence (usually: a homomorphism) between them. Thus, it may be the case that in the transition from N to N' you discover that $M'_{pp} \neq M_{pp}$ in the rather trivial sense that the terms used to describe the elements of M'_{pp} are not the same as those used for the elements of M_{pp}. But as long as you can plausibly construct a correlation between the respective structures, this is all you need to be able to

---

[13] To formulate (*c*) and (*d*) we have assumed that $\bar{I}_N$ and $\bar{I}_{N'}$ are finite sets so that we may numerically compare them. This is a natural assumption if we take into account that intended applications are actually intended by some concrete people at concrete times. However, we could also make use of a more abstract notion of $\bar{I}_N$, which wouldn't exclude an infinite number of structures. Under the latter interpretation, we should modify the formulation of (*c*) and (*d*) but I don't think this is the place to discuss such technical problems at length.

speak of "the same" M_{pp} and to define the conditions of (*PI*) correspondingly[14].

Keeping this in mind, it seems clear that most purported historical examples of paradigm-shifts accompanied by incommensurabilities are of the (*PI*), or even just of the (*SPI*), sort. Take Ptolemy and Copernicus: their respective theoretical assumptions differed radically; and perhaps they didn't have exactly the same intended applications in mind (though this is a matter of historical controversy). However, what seems pretty clear is that both shared the same way of conceiving their intended applications: celestial bodies moving on epicycles. And the reason for this, in turn, is that they shared the same underlying theories: essentially, Babylonian astronomy plus Appolonius' geometry. Another example: consider the transition from caloric theory to thermodynamics. The theoretical terms 'caloric' and 'entropy' may be as incommensurable as you wish. But Laplace and Clausius shared the same notions of 'pressure' and 'volume', and applied them to essentially the same systems.

It may not be an easy matter at all to unearth, in a particular historical example, where exactly the borderline between the *T*-theoretical and the *T*-non-theoretical layers runs, and in what terms exactly $\bar{I}_N$ is described. When reconstructing specific historical examples, we should be prepared to encounter surprises. But from the fact that a particular distinction is sometimes difficult to make in precise terms it does not follow that it is meaningless or hopeless. The adequate (metatheoretical) interpretation of historical data need not be easier for the history of science than it is for the other domains of history. As far as I can see, it is plausible to assume that in most cases of dramatic theory-change mentioned in the literature, there is a common underlying layer of shared notions not questioned by the rivalling theories. It may not be a trivial task to find out exactly where this layer lies but that is a different matter.

Of course, the standard epistemic relativist won't like the distinction between a *T*-theoretical and a *T*-non-theoretical layer. He/she will object that "all concepts are theory-laden". Granted; but the relativist has to explain to us more carefully what he/she means by this theory-ladenness: "theory-laden" with *which* theory? If the relativist wants to say that all scientific concepts are theoretical with respect to one and the same big theory embracing all reality, then this is just a piece of speculative, pseudo-Hegelian bluff we can forget about. But the rela-

---

14 More on the technical aspects of such a construction may be found in (Balzer *et al.* 1987, Ch. VI.5).

tivist may want to say one of two other things which we should examine more carefully.

One possible, serious way to understand "theory-ladenness" is that all concepts appearing in *T* are laden with *some* theory – be it *T* itself or some other theory *T'*. Structuralists are not sure that this claim is true for all conceivable theories but they are convinced that it is true for the vast majority of them. They would just point out that, among the concepts of *T*, one should distinguish those laden with *T* itself (in a sense still to be specified) from those laden with a $T' \neq T$. The two-layers hypothesis about the conceptual structure of theories is compatible with this sort of theory-ladenness.

Another, more radical, way to understand "theory-ladenness" is the claim that, for any *T*, *all* concepts appearing in *T* are laden with *T* itself. This claim, in turn, may be understood in either one of two possible ways: 1) the meaning of all concepts of *T* is at least partially determined by *T*; 2) for all concepts of *T*, the question whether they apply or not to a given situation is *always* decided by *T*, and *T alone*. Now, reading 1) is plausible, whereas reading 2) is not. 1) is plausible because the (Fregean) sense of a scientific term is at least partially conveyed by the net of law-like relationships where the term appears, and such relationships, in turn, rest, at least in part, on the axioms of *T*. (When *all of them* rest on the axioms of *T*, structuralists say that the term is *T*-theoretical.)

On the other hand, reading 2) is *not* plausible. The reason is that scientific theories (perhaps in contradistinction to metaphysical theories) are neither semantically nor methodologically closed units. Science is not a system of self sufficient monads; it is rather a system of checks and balances. A part of theory *T* may be used to find out when a different theory *T'* has gone astray; this part is *T'*-non-theoretical. That science doesn't consist of semantic and methodological monads is not a logical truth; but it is as obvious as any (meta-)empirical matter of fact may be.

This short discussion about theory-ladenness enables us to conclude the following important thing for our topic: we can have (some reasonable sort of) theory-ladenness accompanied by theoretical incommensurability, and at the same time epistemologically relevant comparability at the level of the domains of intended applications. Success and failure when trying to apply *K* to *I* may be comparable across the most dramatic scientific revolution and in spite of theoretical incommensurability and theory-ladenness. And this is all we need for genuinely scientific progress.

8. Progress with total incommensurability?

The hard-nosed epistemic relativist may still be dissatisfied with the analysis. S/he may object that the "real" incommensurability s/he is interested in and which s/he takes to be characteristic of scientific revolutions is not partial but *total* incommensurability – not only incommensurability with respect to the cores but also with respect to the domains of intended applications. S/he may contend that in the really important paradigm-shifts not only K and K' but also I and I' are semantically and methodologically incomparable – there is no common basis to allow for a comparison.

There are several remarks to make to the foregoing contention. In the first place, it is far from obvious that there is total incommensurability in the historical examples typically adduced by the epistemic relativist. Quite the opposite of his/her contention seems to be the case. For obvious reasons, this is not the place to substantiate my historical counter-argument in detail. However, let me provide some hints at the typical historical examples of the incommensurabilists. It seems fairly clear that *there are* shared intended applications along the line Ptolemy-Copernicus-Kepler-Newton-Einstein (even if the *terms* in which they are described may differ), e.g. Mercury's observable path; *there are* shared intended applications along the line Stahl-Priestley-Lavoisier-Dalton, e.g. endurable transformations of metals when exposed to the air; *there are* shared intended applications along the line Buffon-Lamarck-Cuvier-Darwin, e.g. anatomical differences and similarities between fossils and present-day organisms... There may be some difficulties in the precise reconstruction of the conceptual structure of these shared applications; but their actual existence can hardly be denied, unless we wish to play the game of the frivolous snob. Consequently, in these examples and many other of a similar nature which have been brought out in the literature on incommensurability, there is every reason to assume that a common basis of shared intended applications can be detected and, with respect to its common basis, genuinely scientific progress may be defined in the manner of sections 6 or 7 above.

However, let's examine the problem from a more general perspective. Let's consider what the conceptual structure of a total incommensurability would look like. In such a case, not only K_N and $K_{N'}$ but also I_N and $I_{N'}$ would be semantically incomparable. That is, the terms employed to describe $I_{N'}$ and I_N would be such that they are not only different but the semantic components (both referential and non-referential) of their meanings don't have any plausible, specifiable correlation.

Now, the first thing to note about a would-be situation like this is that one can certainly produce an indefinite number of actual historical examples of this sort but they are not relevant at all for the issue of scientific progress. Take this example: N = the caloric theory; N' = the Marxian theory of value; N and N' clearly succeeded in time (the first died out more or less at the time when the second emerged); and clearly, not only K_N and $K_{N'}$ but also I_N and $I_{N'}$ are completely incommensurable from a relevant semantical point of view. So, we have here all the conditions fulfilled that the relativist apparently thinks of. Should we infer from the historical reality of this example anything about the possibility or impossibility of genuinely scientific progress? Of course not. Nobody, neither the progressivist nor the relativist, nor anybody else would have the idea of asking whether there was progress in the "transition" from the caloric theory to the Marxian theory of value. Everybody (including the most stubborn relativist, I suppose) agrees that this example is completely irrelevant for the question of scientific progress. Note that the reason for this is not that the two theories have *nothing* in common: as a matter of fact, they have a lot of interesting things in common: e.g. both were produced in Europe, both emerged after the Industrial Revolution, both were devised by men of a materialistic outlook, etc. So, the reason for the felt triviality of the example is not the lack of some common aspects, but the lack of relevant common factors for scientific progress.

Consequently, the epistemic relativist is confronted with the following dilemma. Either s/he wants to claim that all interesting and real cases of "scientific revolutions" or "paradigm-shifts" have the same semantical structure as the caloric theory – Marxian value theory example – a claim nobody would reasonably believe. Or else, s/he wants to claim that the interesting cases of incommensurable theories have some non-trivial common basis for comparison. But, then, the *weakest reasonable* common basis I can think of for such a comparison is a shared domain of intended applications, or at least mutually comparable domains of intended applications, be they technological or of some other kind. Less than this leads to the caloric theory – Marxian value theory sort of example. But this weakest reasonable common basis is already enough, as we have seen, to define a plausible concept of genuinely scientific progress and produce a great number of historical instantiations of it.

Institute for Philosophy, Logic and Philosophy of Science
University of Munich

REFERENCES

Balzer, W., Moulines, C.U. and Sneed, J.D. (1987). *An Architectonic for Science*. Dordrecht: Reidel.

Diederich, W., and Ibarra, A. and Mormann, Th. (1994). Bibliography of Structuralism. *Erkenntnis* **41**, 403-418.

Feyerabend, P.K. (1977). Changing Patterns of Reconstruction. *British Journal for the Philosophy of Science* **28**, 351-368.

Hacking, I. (1983). *Representing and Intervening*. Cambridge etc.: Cambridge University Press.

Kuhn, T.S. (1962, 2nd ed. 1970). *The Structure of Scientific Revolutions*. Chicago: University of Chicago Press.

Kuhn, T.S. (1976). Theory-Change as Structure-Change. Comments on the Sneed Formalism. *Erkenntnis* **10**, 179-199.

Moulines, C.U. (1991). *Pluralidad y recursión* [Plurality and Recursion]. Madrid: Alianza.

Moulines, C.U. (1991). Über die semantische Explikation und Begründung des wissenschaftlichen Realismus. *Dialektik* **1**, 163-178.

Stegmüller, W. (1973). *Theorienstrukturen und Theoriendynamik*. Berlin: Springer etc. (English translation by R. Wohlhüter, *The Structure and Dynamics of Theories*. New York: Springer 1976).

Poznań Studies in the Philosophy of the Sciences and the Humanities
2000, Vol 72, pp. 199-234

Adam Jonkisz

ON RELATIVE PROGRESS IN SCIENCE[1]

ABSTRACT. A concept of relative progress in science is given and justified against a background of other conceptions of progress. Using the methods of the structuralist view enables one to achieve a group of new concepts which make it possible to characterize the magnitude of scientific changes and progress, differentiate between kinds of progressiveness and determine its degree and effectiveness. The results are the basis for a quantitative characterization of progress and for the use of some geometrical methods to solve related problems (e.g. the definition of the direction of changes and the quantitative relation between magnitude, direction and progressiveness of a given scientific change).

1. General remarks on progress in science

The general notion of progress has many types of designata; it also assumes many simpler notions. By limiting our considerations to progress in science, we exclude a number of domains in which we could identify progress, but such a limitation is still not sufficient to determine unequivocally the scope of the term "progress". In the realm of the scientific, we could distinguish the so-called background of science and then discuss progress in terms of improving research techniques (progress in organization and methods) or, more narrowly, in terms of improving measuring techniques (instrumental progress), or even merely follow the growth of academic institutions (institutional progress). It could even be possible to apply some numeric criteria of progress in that domain, e.g. ones based on the accuracy of measuring instruments expressed in numbers, on the number of active academic institutions, the size of their staff, on the number of journal publications, etc.

A further limitation – to proper science utilizing the above-mentioned background – does not remove the ambiguity either, because the property of being progressive can be ascribed both to scientific activities and to their results. Limiting ourselves to basic results of the scien-

---

1 The present article employs results published in (Jonkisz 1990). Also, in many places it refers to a paper by C. U. Moulines "Is There Genuinely Scientific Progress?" contained in this volume.

tific process still leaves us with both problems and solutions of various kinds. For instance, in the sphere of solutions, we distinguish the establishment of facts from the discovery of regularities in phenomena, and from the formulation of explanatory theories. Theories, in turn, can be of different magnitude, ranging from a single hypothesis to a vast and detailed theoretical system with many consequences in the realm of problems and methods of inquiry. Therefore, even in science conceived in a narrow sense, we can distinguish progress in the following domains: of problems, of concepts, and of explications; we can also discuss progress in empirical and theoretical realms, as we can judge the speed and magnitude of all types of progress[2].

To narrow down the general notion of progress to a notion relevant to our inquiry, we must first distinguish development from scientific progress. We take the word "progress" in the sense of progressive changes being a subset of the totality of changes making up the historical process called the "development of science". Further, we distinguish progress in science itself from progress achieved thanks to its results. Progress in other domains, e.g. in technology, culture, or in civilization, is a result of science, in the sense that without science it would have been impossible. It also happens that improvements in science are fundamental to progress outside of science (e.g. in technology). We are interested in progress in science itself, or more accurately, in the kind of progress called epistemic progress[3].

---

2 Solutions reached in the domain of progress are closely connected to other questions. The most important of them are the goal and the function of science, and questions of comparability, continuity and reduction between theories. There are also more basic problems, such as the question of the cognitive capacities of man, of the object of knowledge, and more specific ones – like those concerning simplicity and empirical confirmation of theories. At the same time, progress is possibly the most general and the most basic concept to be used when making value-judgements both at different stages of the process of acquiring knowledge and in the evaluation of results of science occurring outside science. Calling something progressive usually amounts to acknowledging the rationality of scientific activities and to confirming the importance of all features regarded as essential to science because of the goal ascribed to it. Many of those problems can be dealt with independently of any conception of progress, but the problem of progress cannot be tackled at all without accepting some assumptions, at least in the domains of comparability and continuity, and without accepting some solutions regarding the goal and functions of science.

3 A full and systematic overview of conceptions of epistemic progress can be found in (Hajduk 1989/1990 and 1991/1992). The name might suggest identifying progressiveness with the achievement of better and better knowledge of reality, thus favoring conceptions of science ascribing this goal to science. Even if one accepts such an identification as irreversible, it is still better to discuss progress with respect to the goal *ascribed* to science to avoid suggesting anything.

That questions regarding epistemic progress are sufficiently justified is revealed as a result of doubts concerning the understanding of "knowledge", particularly if we are looking for effective criteria of progress of our knowledge. To say that knowledge grows with the development of science seems a commonplace, just like saying that increasing knowledge is equivalent to decreasing ignorance. These beliefs are illustrated by Fig. 1: the area of knowledge expands at the expense of the area of ignorance; there are fewer and fewer questions to answer.

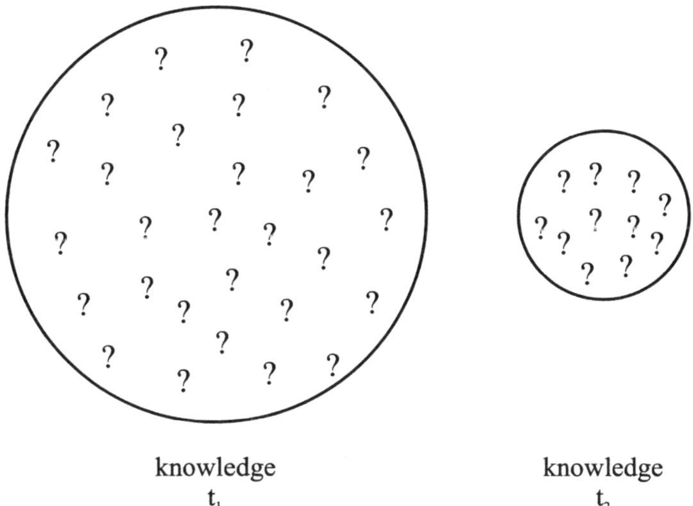

Figure 1

However, it is by no means sure that knowledge grows (how do we understand knowledge: as probable?, true?, as concerning what?) or that it does so at the expense of ignorance. If we want to eschew doubts about the growth of knowledge, then we have to treat the expanding circle as an ever-increasing set of questions that are no longer valid and of answers usually corresponding to them. Problems go out of date either because they are solved (they cease to be sound) or because they are demonstrated to be inaccurate (to rest on false assumptions), or because they are shown to be insoluble. It is known that properties of questions determining whether they are valid (or more generally: well-posed) depend on the state of knowledge: on accepted statements, conceptions and especially on systems of concepts and statements called theories. But in this case, there is nothing like an area of ignorance "waiting" to be eradicated by growing knowledge. Ignorance exists only in the sense of valid questions, questions rooted in theories being in force. Consequently, it is easy to draw the conclusion that knowledge and ignorance (understood as above) grow simultaneously: ignorance is generated by knowledge (Fig. 2).

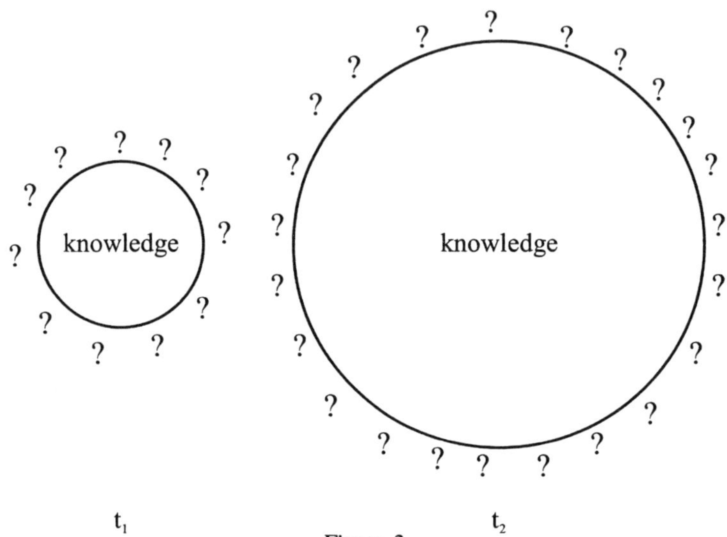

Figure 2

In the area of science taken in general, we can distinguish parts corresponding to disciplines of science, or we can trace even finer developmental trends. The degree to which we consider science to be an activity or to be a set of results of an activity can also vary. We are interested in progress – as distinct from development – within the physical sciences and we consider it from the perspective of the results of scientific activities, i.e. from the perspective of physical theories. In short, we are interested in the progressiveness of physical theories.

2. Progressiveness of physical theories

We shall investigate the problem of the progressiveness of theories in the context provided by the so-called structuralist view (J. Sneed, W. Stegmüller, W. Balzer, C.U. Moulines, and others). Our considerations assume, first of all, the structuralist method of reconstructing physical theories. That view represents theories as so-called theory-nets. A single theory-net N stands for a single version of a theory, so the development of the theory must be represented as a series of appropriate theory-nets (N_i: $i = 1, ..., n$). A theory-net is an ordered set of so-called theory-elements. Every theory-element T consists of a particular mathematical structure K called the theory-core and of the set I of so-called intended applications ($T = \langle K, I \rangle$). In the classical reconstruction, the theory-element corresponds to a coherent set of laws of a theory (K) applied in a subset of the totality of empirical domains in which the theory is applied. The relation ordering elements in the net N is the so-called specialization (σ): the notation "$T' \sigma T$" read as "the theory-element T' is a specialization of the element T", which means that the theory-core K' is a specialization of the core K ($K' \sigma^* K$) applied in a narrower scope of intended domains ($I' \subseteq I$).

If a given version of a theory is understood strictly, i.e. as applied by a particular scientist at a particular time, then it is possible to show that the net representing the version of the theory has a tree-like structure (in the set-theoretic sense). At the top of the net N representing the version of the theory is the so-called basic element $T_0 = \langle K_0, I_0 \rangle$, whose specializations are all the remaining elements. The set of all generalizations T' of any element T in a given net (in the sense: $T \sigma T'$) forms a chain, which means that T has in N only one direct generalization. This ordering of the net N is preserved in its mereological parts, i.e. in the so-called core-net (N^*) containing cores of theory-elements from N ordered by the relation σ^*, and in the net called the application-net (N^+), in which extensions of those elements are ordered by the relation of inclusion $\subseteq$[4].

In the structuralist view, components of the core K are described in detail. For our purposes, however, it suffices to state that K comprises, among other things, the set M_p of so-called potential models, the set M of models called real or proper models (where $M \subseteq M_p$) and the set M_{pp} of so-called partial potential models. The set of partial potential models contains so-called non-theoretical reducts of elements of the set M_p ($M_{pp} = r(M_p)$). Determining the set M_{pp} requires a theory-related criterion of theoreticity of concepts: elements of M_{pp} are the domains which can be described independently of concepts (functions) defined on the grounds of a given theory, i.e. using only concepts whose meanings have been determined in prior theories. Such domains were called empirical domains above, but it is more appropriate to say that a domain is empirical relatively to a given theory. The non-theoretical level un-

4 An approach which makes it possible to prove the claim about such a structure of all empirical theories is presented in (Jonkisz 1990). The interpretation of the version of the theory accepted in (Balzer *et al.* 1987) does not permit the acceptance of the claim as generally valid.

derstood in this way comprises of the set M_{pp} and of the set I of intended applications ($I \subseteq M_{pp}$). The theoretical level, in turn, contains the set of potential models M_p and the set M of models ($M \subseteq M_p$). Physicists' belief that a given theory is applicable in the domain $y \in I$ can be expressed briefly as $y \in M$. The totality of such beliefs concerning a certain theory-element of a given theory are contained in the statement: $I \subseteq r(M)$, also called the empirical claim of the element T. An analogous statement containing extensions of all theory-elements of the net N in its scope is called an empirical claim of a theory. Empirical claims of particular theory-elements and the empirical claim of a theory change as a result of changes in the theoretical (N^*) or empirical (N^+) parts of the theory.[5]

Before we start looking for solutions to the problem of the progressiveness of theories, it is desirable to state clearly what we demand from a good theory of progress. Hence, a full conception of progress answers the question of what progress consists in, as well as determines an effective means of its discrimination. A conception of progressiveness should, then, not only explain how we ought to understand progress which comes about a result of changes in theories, but should also equip us with a criterion of progressiveness.

Further, we want a satisfactory conception of progress to define its own concepts and tests irrespective of the division of changes in science into intra-theoretical (i.e. within one theory) and inter-theoretical. The distinction between the "evolutionary" development of one theory and the "revolutionary" transition to a new one, common in philosophy of science, often goes hand in hand with the belief that fundamental diachronic problems (of comparability, continuity, reduction, or progress) become meaningful only in the case of transitions to new theories, and that they are easy to resolve, or maybe even not valid, in the context of evolutionary changes. Yet, contrary to that practice and to that belief, we contend that we ought to strive to obtain one coherent set of concepts applicable in both contexts, to have an understanding and a criterion of progress that can be applied successfully regardless of whether we want to investigate progress in the development of one theory or progress realized as a result of replacement of theories. Therefore, the concept of transition m, $m = \langle N, N' \rangle$, used in investigating the progressive-

[5] To facilitate reading and comparing texts contained in this volume, I adopted simplifications and abbreviations suggested by C. Moulines in "Is There Genuinely...," to which I refer the reader at many points in my text. I leave out explanations of basic concepts of the structuralist view given there and references to works explaining in more detail those concepts and related ones.

ness of theories ought not be limited to development of one theory: N and N' are particular versions of a theory, but they can stand both for two phases in the development of one theory and for versions of two different theories. Transition from N to N' in the general sense is, then, transition in the development of a discipline[6].

Solutions to the problem of progress advocated by prominent structuralists can be reduced to one intuition, usually hidden behind elaborate formalism: progress is achieved when a new theory (or version of a theory) adds its own results while preserving existing ones[7]. In the context of one theory, that means formulating new specific laws or broadening the range of applications to comprise new domains. Thus, cases of progress coincide with growth and cases of regress coincide with decrease, i.e. with rejecting hitherto accepted laws or with excluding some domains from the range of applications of the theory. Tests of progress and regress are also straightforward: they are based on the relation of set inclusion[8]. If one considers transition to a new theory, then the relation representing progress must be determined in a way

---

[6] It turned out that satisfactory criteria of identity of theories – always assumed during discussions of such divisions – can be formulated only under such a wider approach, i.e. in the context of the development of a whole discipline. Cf. (Jonkisz 1990, pp. 81-118, 376-386), where the necessity of such an approach is justified and where criteria making improvements to earlier solutions to the problem of identity of theories are given.

[7] Cf. (Sneed 1971, pp. 216-234), (Sneed 1977, p. 128), (Balzer *et al.* 1977/78, pp. 202-204), (Balzer *et al.* 1987, pp. 275-284). This basic intuition had already been formulated in research on progress; cf., e.g. (Lakatos 1970, pp. 116-119, 155).

[8] Cf. (Sneed 1971, pp. 247-307), (Stegmüller 1976, pp. 220-223), (Stegmüller 1978, p. 54). A slightly different solution is proposed in one of the earlier works by C. Moulines, in which progress is identified with instances of empirical growth and stagnation: a theory develops in a progressive way if the class of its successful applications does not decrease. Moulines also defines the concept of perfect evolution, or of growth as a result of which all earlier postulated applications of a theory are confirmed; cf. (Moulines 1979, pp. 425-426). Growth in the two strata is also a necessary criterion of the so-called progress in Lakatos' sense, which C. Moulines formulates in his article contained in this volume. Incidentally, let us notice that the condition (a) of that formulation ought to be narrowed down to cores ($\forall K_i' \, \exists K_j \, (K_i' \, \sigma^* \, K_j)$) because this condition required of theory-elements is not appropriate: the transition $m = \langle \{T\}, \{T'\} \rangle$ such that $K' \sigma^* K$ and $I \subset I'$ is, without doubt, a case of progress in Lakatos' sense, but it does not satisfy the condition (a) in its present formulation. On the other hand, accepting the condition (b) as being part of the definition of the *criterion* of progress understood in such a way requires the acceptance of a test that makes it possible to decide for any domain of the intended scope of a theory whether it is a (partial) model of the theory (whether it belongs to the class $I \cap r(M)$).

that makes it possible to connect: (i) theories having different sets of concepts and laws and whose ranges of applications cannot be compared by means of the relation of set inclusion; but (ii) only such pairs of theories that realize the transfer of results. These requirements are satisfied by the relation of reduction ρ, which is called, after Stegmüller, macrological reduction, in order to distinguish it from the traditional understanding of reduction. It is maintained that progress is achieved if and only if the rejected theory reduces to the new one[9].

It is possible to provide reasons why conceptions based on the idea of cumulativism are not right or, at least, not sufficiently general. This objection is valid no matter what version of cumulativism one chooses to discuss and no matter what particular definition of progress one adopts. All those conceptions neglect an important property of progress, i.e. its relativity. The kind of relativization we discussed above – to a given domain or to an aspect – is obvious. But that is not essen-

---

[9] Cf. (Sneed 1976, p. 136), (Stegmüller 1975, pp. 96-97), (Stegmüller 1976, p. 227), (Stegmüller 1978, pp. 58-63), (Balzer *et al.* 1977/78). Corrections made to newer definitions of the relation ρ do not change the fundamental idea of reduction; cf. (Balzer *et al.* 1978, pp. 252-321, DVI-5). Conditions defining the relation of reduction can also be found in the latest criterion of progress formulated by C. Moulines for cases of exchange of incommensurable theories (cf. his paper in the present volume). If one ignores conditions regarding elements of the core left out by Moulines (i.e. the conditions DVI-5(4) and DVI-5(5)), then one can see that the remaining ones are taken into account in the definition (*SPI*) and in the condition (b) of the definition (*PI*). One just has to remember that the sets S occurring in those definitions contain intended applications in which laws of the theory ($S = (I \cap r(M))$) are satisfied.

To recognize properly the version of cumulativism contained in the concept of macrological reduction, let us distinguish three versions of cumulativism: cumulativism connected with content, with domains, and with functions of scientific theories. According to the first version of cumulativism, there is accumulation in content of theories, in the sense that prior statements belong to the set of statements of a subsequent theory. Cumulativism concerning domains boils down to the belief that there is growth (again, revealed by means of the relation of inclusion) of solved problems or explained phenomena (i.e. problems that can be formulated and phenomena that can be described independently of concepts of both juxtaposed theories). The third version of cumulativism only deals with a new theory taking over the basic functions served by its predecessor (as explanation, prediction etc.). The versions of cumulativism were listed from the strongest to the weakest: from the thesis of accumulation of content it follows that there is growth in domains and in functions, and from the thesis of accumulation of domains it follows that functions are taken over. But converse relations do not obtain. The thesis of cumulativism contained in the concept of reduction belongs to the version connected with domains because it requires that phenomena explained by a prior theory be found in the domain of phenomena explained by the reducing theory.

tial here. Of fundamental significance is the unavoidable, although perhaps overlooked or concealed, reference to a pattern which forms the basis of judging changes with respect to their progressiveness. This relativity is unavoidable in every domain in which progress is proclaimed, but it is more obvious in the social realm than in science. If we want to find this kind of relativization in scientific progress, let us just think about judgements issued on transitions whose achievements were consequently rejected. Are we really going to take as progressive any increase in the number of laws or known domains of applications of a theory regardless of whether the proposed innovations have been preserved or rejected as a result of further research? How are we going to judge the elimination of such inappropriate components (mistaken laws, statements about the range of applications, whole theories based on wrong assumptions concerning physical objects and processes)? Is growth always a sign of progress; does elimination and decrease always signify regress?[9] In the context of such questions, we realize a need to formulate a conception of progress that would take into account not only the idea of cumulativism but also the irremovable relativity of progress.

3. Relative progress

The conception of relative progress offers the possibility of solving the above problems. We have to acknowledge that progress must always be related to the point from which it is evaluated. In science, too, one judges as progressive only that which has pushed development towards accepted scientific results, the most important of which are accepted theories. This kind of evaluation can be seen, for instance, when a historian of science notices valuable components or ideas in old, long since rejected conceptions. Past scientific results gain or lose value with respect to present theories, or more generally, with respect to the pattern of science in whose light historical facts are described.

The conception of relative progress explains why the very same changes in science can be (justly) evaluated as progressive or to the contrary – as regressive. It explains why evaluations issued by authors of changes run a great risk of being questioned, and why such evalua-

[9] A good example of such a dead end, which scientists ought to have avoided, as became clear from a later perspective, was a way that seemed to lead to a theory of dynamic Universe by analyses of the transformations of de Sitter's world metric. Cf. (Turek 1996).

tions must always be favorable. Divergence of evaluations results from the fact that they are built on the basis of different patterns of science. Additionally, the probability of a future change of results is greatest in the case of transitions occurring at present. But if one accepts the results of such transitions, then they create a new basis for evaluations. A scientist being in favour of a new theory, *T*, always pronounces the progress achieved by the advent of the new theory because he judges progress from the standpoint of that new theory (it is obvious that transition to the theory *T* pushes the development of science towards the pattern whose essential component *T* itself is).

The proposed understanding of progress is independent of the structuralist view. However, the methods of set theory used in this approach make it possible to formulate precisely and to justify statements concerning progress, and to provide the means of its verification and comparison[11].

4. Recognizing progress

The conception of relative progress captures two important components of progressiveness. It forces one to determine whether an earlier state of knowledge improved; it also requires an assessment of whether improvements have taken place in the right direction. It is even possible to distinguish two kinds of progress: progressiveness in a particular transition (contextual) and progressiveness in the development of a discipline (resultant or historical). Effective recognition of progressiveness of the first kind requires a criterion enabling one to decide in each case of transition whether the final state is an improvement over the original state. Resultant progressiveness is proclaimed from a perspective going beyond a given transition (excluding special cases in which the pattern of evaluation is identical to the final state of the transition), and it requires the right direction of changes to be determined.

4.1 Contextual progressiveness

The easiest way to describe contextual progressiveness in terms of set theory is to use the relation of inclusion. The effectiveness of this test of progress depends on whether compared stages of science are always (if not always, then how often) comparable with respect to the relation of set inclusion. A brief moment of thought leads us to the conclusion that it

---

[11] Cf. (Jonkisz 1984), (Jonkisz 1990, pp. 273-284).

might not be so even in the development of one theory. A trivial example of such incomparability with respect to the relation $\subset$ is the transition $m = \langle\{T\}, \{T'\}\rangle$ such that $K \neq K'$ and $K' \sigma^* K, I \subset I'$, i.e. the transition from a more general theory with less content to a more specific one, describing a wider range of phenomena more accurately. The transition m is a clear example of contextual progress but a test based on the relation $\subset$ would force us to adopt a contrary evaluation. Within our general approach, in which the context of one transition, $m = \langle N, N'\rangle$, can exceed the development of one theory, the limitation to the relation of inclusion would have much worse consequences. It is so, because in the case of a theory change, the chances of obtaining a simple inclusion are much smaller. On the other hand, many transitions, especially in the development of one theory, can be correctly described using the concept of inclusion.

A better solution is based on the very simple relation *in* binding any two theory-elements $T = \langle K, I\rangle$, $T' = \langle K', I'\rangle$ if and only if $K' \sigma^* K$ and $I \subseteq I'$. If it holds that T *in* T' in the sense we have just described, then we can say that T is embedded or is contained in T'. Similarly, we can say that the theory-core K is embedded or contained in K' (K $in^*$$K'$). The concept of containment allows us to ascertain implicit presence of an element in a theory-net, so it gives one a basis to generalize the simple relation of belonging to a set and simple inclusion of sets of theory-elements.

(D1) The theory-net N is contained in the net N' ($N \prec_{\text{in}} N'$) *iff* every element T of the net N has its counterpart T' in the net N' in which it is embedded (in the sense of T *in* T').

Containment in the sense of **D1** obtains obviously if $N \subseteq N'$. It is often the case, though, that it is necessary to develop a net richer than N' until N is contained in it in the sense of $\subseteq$. We can now distinguish the set $ex(T)$, which is an exposition of the element T ($ex(T) = \{T'\colon T'$ *in* $T\}$), and various expositions of the original net obtained by adding or revealing of components embedded in (*in*) its elements. A banal example of such an exposition needed in the above situation of transition from the net $\{T\}$ to the net $\{T'\}$ is the net $N'' = \{T', T''\}$, whose second element is such that $K'' = K$ and $I'' = I$. We can see that T'' *in* T' and $\{T\} \subseteq N''$, which enables us to conclude that $\{T\} \prec_{\text{in}} \{T'\}$[12].

[12] It is easy to find versions of the relations *in* and $\prec_{\text{in}}$ referring to theoretical or empirical sides of the process of development, as well as many derivative relations; cf. (Jonkisz 1990, **D2**, **D6**, **D7**, pp. 62-64, 100-102). All relations among theory-elements

The test of contextual progress based on the relation $\prec_{in}$ enables us to take into account all situations in which elements of a prior net are embedded in components of the subsequent net, i.e. are contained in it *implicitly*. This solution is better, but it is still not general enough. The relation $\prec_{in}$ between nets is ultimately based on the relation of specialization σ, which binds theory-elements of significant structural and functional similarities, that is, elements based on common theoretical assumptions (laws and concepts) used in domains compared by means of the relation of inclusion. This relation is suitable only to reveal connections among elements or versions of one and the same theory. In our approach, in which nets making up the transition $m = \langle N, N' \rangle$ can stand for versions of different theories, a more general relation is needed.

The concept of macrological reduction gives us the possibility of discovering further situations in which contextual progress takes place. We have already established that this concept contains the idea of improvement, which consists in a new theory (version of a theory) preserving all results of the preceding theory while adding its own ones. It is easy to see that the test of contextual progress based on the reduction relation ρ is a generalization of tests based on the relations $\subseteq$ and $\prec_{in}$. When we look at single elements, we can notice a generalization, proceeding from the relation $=$ to *in*, and then from *in* to ρ: if $T = T'$ then T *in* T'; if T *in* T', then $T \rho T'$[13]. These basic intuitions and relations justify considering the relation of reduction, but we still have to learn more about this relationship and about its applications to understanding progress[14].

Scientific theories (or more accurately, their specific versions) are represented by theory-nets, so it seems obvious that reduction should be defined as a relation between sets of theory-elements. The evidence of progress would then be reduction between nets. Such a solution would be correct if one identified progress with contextual progressiveness. Reduction of a prior net to the subsequent one correctly points to

---

and among theory-nets order sets of theory-elements and of theory-nets, respectively (*ibid*, **T1A-B**, **T2A-C**, **T11A-B**, pp. 64-66, 104; and 366-372, 380-384 where proofs are given)

[13] Cf. (Jonkisz 1990, **T41**, pp. 288; 447)

[14] In (Jonkisz 1990) I used the definition of the relation of reduction proposed in (Balzer *et al.* 1987, p. 227, DVI-5, DVI-6), but one can also accept other definitions. Let us notice that both earlier and present analyses admit a more general type of reduction, which allows the mediating element T^* being a specialization of the reducing element T'. The need for such a generalization is demonstrated by examples of reduction between physical theories given in (Balzer *et al.* 1987). Those examples show that the condition of direct reduction is not always satisfied (*ibid,* pp. 259, 262, 273, 277).

an improvement of the original state of knowledge, but it does not take into account the direction in which the transition leads. Can we regard as progress a situation where a reducing theory is subsequently abandoned? Is progress realized when a whole progressive branch – i.e. a series of theories or of versions of one theory exhibiting contextual progress (in which previous theories are reduced to subsequent ones) – is cut off as a result of further development?

The relation of reduction itself, no matter how well defined, does not suffice to describe progress in a satisfactory way. Proclaiming progress assumes more than the concept of contextual progress.

4.2 Resultant progressiveness

Resultant, or – in other words – historical, progressiveness is accounted for by means of the so-called net of results or, speaking more accurately, the net N_r of results accepted within a discipline. The concept of the net of results is a pragmatic one. It refers not only to the practice of natural sciences but also to the practice of metasciences whose subject of study they are. The net of results can be made up of any set of outcomes achieved in the development of a discipline, if it is accepted either by physicists themselves or by "metaphysicists" reconstructing the development. The set can come from the development of one theory or it can contain outcomes of different theories; it can even comprise the totality of theories accepted at a given moment. The scope of the set depends on what (and how large) a fragment of a discipline one wants to cover. We mean, then, the totality of results taken as a basis for evaluating changes in terms of progress. The general idea of such an evaluation is simple: the only progressive changes are those that lead towards results contained in the net N_w.

Such an understanding of the net of results suggests the rejection of relations defined over whole nets in investigating progress. Such relations, e.g. reduction between nets, are too synthetic. During evaluation of a certain transition with respect to progress, it may turn out that not all elements of the final net, judged from the perspective of an accepted net of results, are correct. We could say briefly that not all elements exhibit resultant progressiveness. An effective conception of progress ought to enable one to distinguish components with resultant progressiveness from others. Reduction between nets treats theories represented by theory-nets globally. Consequently, it does not enable one to go down to the level of components of a theory and it precludes such distinctions. Therefore, in our description of progress we make use of the relation of reduction defined on single elements.

A general approach to progress requires one to properly distinguish elements of a given transition $m = \langle N, N' \rangle$ according to two dichotomic principles: into elements that are retained and rejected, and into progressive and regressive ones.

(D2) If $m = \langle N, N' \rangle$ is a transition and N_w is a certain net of results in the development of a discipline, then $T \in (N \cup N')$ is:

A. an element retained in m *iff* there is a $T' \in N'$: $T \rho T'$; if such an element does not exist, then T is an element rejected in m.

B. an element progressive with respect to N_w *iff* there is a $T_w \in N_w$: $T \rho T_w$; if such an element does not exist, then T is a regressive element with respect to the net N_w.

The former division is made within a given transition and it has to do with contextual progressiveness. Separating progressive and regressive components requires going beyond the transition m – to the net of results N_w, and it serves the purpose of capturing resultant progressiveness. In such a case, one looks for elements of the net N_w corresponding to a given element of T: if there exists a component of the net of results to which T can be reduced, then we call T progressive, otherwise – regressive.

These distinctions show that any transition evaluated in this way can have both progressive and regressive dimensions. The two complementary aspects can be described quite precisely.

(D3) If $m = \langle N, N' \rangle$ is a transition and N_w is a particular net of results in development of a discipline, then:

$m_{pr} = \langle N_{pr}, N'_{pr} \rangle$ is a progressive aspect of the transition m *iff*

$N'_{pr} = \{T \in N'$: T is a progressive element with respect to $N_w\}$

$N_{pr} = (N^P_{pr} \cup N^N_{pr})$ *where*

$N^P_{pr} = \{T \in N$: T is an element retained in m and progressive with respect to $N_w\}$

$N^N_{pr} = \{T \in N$: T is an element rejected in m and regressive with respect to $N_w\}$

$m_{rg} = \langle N_{rg}, N'_{rg} \rangle$ is a regressive aspect of the transition m iff

$N'_{rg} = \{T' \in N'$: T is an element regressive with respect to $N_w\}$

$N_{rg} = (N^P_{rg} \cup N^N_{rg})$ where

$N^P_{rg} = \{T \in N$: T is an element retained in m and regressive with respect to $N_w\}$

$N^N_{rg} = \{T \in N$: T is an element rejected in m and progressive with respect to $N_w\}$.

Both the progressive (m_{pr}) and the regressive (m_{rg}) sides of the transition m are represented by subsets of the totality ($N \cup N'$) of elements occurring in m. In those subsets, we can, in turn, distinguish groups of elements corresponding to more detailed aspects of progress or regress. The subset representing progress is determined according to the following self-evident suggestions. Progress is realized when one retains the correct element (N^P_{pr}), rejects the incorrect one (N^N_{pr}) or adds the correct one (N'_{pr}). Regress consists in retaining incorrect elements (N^P_{rg}), rejecting correct ones (N^N_{rg}) and adding incorrect ones (N'_{rg}). A correct element is an element progressive with respect to the net N_w; a incorrect element is an element regressive with respect to it (in the sense outlined in **D2.B**). It is possible to verify that the above characteristics guarantee that the aspects of transitions do not overlap and that all elements of transitions are accounted for in this division[15].

The concepts of progressive and regressive aspects of transitions allow us to define certain boundary types of transitions: entirely progressive transitions and entirely regressive ones. The former transitions are identical to their progressive aspect (if $m = \langle N, N' \rangle$ then $(N \cup N') = (N_{pr} \cup N'_{pr})$) while the latter are identical to their regressive aspect ($(N \cup N') = (N_{rg} \cup N'_{rg})$). It follows that transitions of the first type have an empty regressive aspect ($m_{rg} = \langle \varnothing, \varnothing \rangle$) and transitions of the second type – the progressive aspect ($m_{pr} = \langle \varnothing, \varnothing \rangle$)[16].

The suggested conception of relative progress enables us to take into account resultant progressiveness and relate it appropriately to contextual progressiveness. It also provides us with a means of separating the progressive and regressive aspects of any transition in the development of any discipline. However, it does not allow us to pass any judgement about progress occurring in transitions taken as wholes. In real transitions, i.e. in those realized in the history of science, changes of both types are often mingled together. Only exceptionally are actual fragments of the development of science purely progressive or purely regressive. Only in those cases do the proposed concepts enable an evaluation of progress of changes taken as wholes to take place. To make such tests of progress effective, it is necessary to have some means of deciding what kind of changes prevails in a given period. That requires a method of comparing the progressive and regressive aspects and thus estimating the total progress achieved in a given transition.

---

15 Cf. (Jonkisz 1990, **W12a-b**, pp. 296-297; 448-449).

16 Cf. (Jonkisz 1990, **D30**, **W13**, pp.297-298; 449).

5. The magnitude of relative progress

A numerical characteristic required to estimate the magnitude of progress assumes, in turn, a means of measuring the size (significance, importance) of changes. If one established the importance of changes occurring as a result of the entire transition *m*, then it would be possible to establish, in the same manner, the size of the progressive and regressive aspects of the transition. The result of comparing the two complementary sides of the transition would yield a basis for assessing the total progress associated with the entire transition. It would be possible to judge not only whether the transition is progressive as a whole, but also to establish the degree of progress or the degree of regress resulting from it.

5.1 Significance of changes

Thus, the most important thing in assessing the magnitude of progress is an effective measure of the significance of changes. We must bear in mind that the measure must be universal if we want to achieve a universal conception of progress. Determining significance of changes must, therefore, be effective regardless of whether *m* is a transition in the development of one theory or a transition from one theory (or a version of it) to another. The so-called universal measure of significance of changes (F_m^{mx}) displays such capabilities[17].

The idea leading to the measure F_m^{mx} is simple. First, one has to present both compared versions of theories in a way that reveals their whole content: not only its explicit components but also all their assumptions, even ones extending beyond the scope of a given theory to physical theories presupposed by it, even – if necessary – to conceptions lying beyond physics. After such a presentation of scientific results making up a given transition $m = \langle N, N' \rangle$, one assesses the size of the transition from N to N', proportionally to the number of components singular to the net N (i.e. not contained in N') and to the number of components specific to N' (i.e. absent in N).

Our earlier remarks on the relations *in* and $\prec_{\text{in}}$ and the above explanations concerning the measure of F_m^{mx} suggest that presentation of the content of compared theories (versions of theories) consists in developing the nets N and N' making up the transition *m*. Because in a general view a need for a far-reaching exposition of nets should be allowed for,

17 Cf. (Jonkisz 1990, pp. 239-252).

it is necessary to refer to so-called maximal expositions in determining the size of a transition.

(D4) If N is a theory-net, then:

A. N^{mx} is the maximal exposition of the net N *iff*
$N^{mx} = \{T'$: there exists an element $T \in N$ such that T *in* $T'\}$;

B. N^{K^*mx} is the K^*-maximal exposition of the net N *iff*
$N^{K^*mx} = \{T'$: there exists an element $T \in N$ such that T' *in* T and $K' \sigma^* K^*\}$;
N^{I^*mx} is the I^*-maximal exposition of the net N *iff*
$N^{I^*mx} = \{T$: there exists an element $T \in N$ such that T' *in* T and $I' \subseteq I^*\}$;
N^{T^*mx} is the T^*-maximal exposition of the net N *iff*
N^{T^*mx} is simultaneously the K^*-maximal and the I^*-maximal exposition of the net N.

The maximal exposition of a given net N is the net containing all the elements embedded in any component of N. The maximal exposition is, therefore, the opposite of the most compact presentation of the content of a given net; it is the most extensive one. This concept is a counterpart of the set of formulae deductively closed, known from the statement view. The kinds of maximal expositions defined in **D4.B** are needed to effectively apply tests based on the universal measure of fundamentality. K^*-maximal, I^*-maximal, and T^*-maximal expositions are subsets of maximal expositions. Limits of expositions are determined by the core K^* (no element appearing in an exposition can have a core more general than K^*), by the scope of I^* (extensions of elements of an exposition must be supersets of the scope I^*), or by a given element T^* (it combines both of these limitations)[18].

Applying maximal expositions in a way enabling us to estimate the size of the transition $m = \langle N, N' \rangle$, lets us define the universal measure of fundamentality of changes F_m^{mx}.

(D5) If $m = \langle N, N' \rangle$ is a transition in the development of a discipline, then $F_m^{mx} = \|(N^{mx} \dot{-} N'^{mx})\|$, where N^{mx} is the maximal exposition of the net N and N'^{mx} is the maximal exposition of the net N'[19].

---

18 Cf. (Jonkisz 1990, pp. 150-153), where other types of maximal expositions are defined and relations among them are unveiled (*ibid*, **D14**, **T17a-b**, pp. 150-153; 397-398).

19 Cf. (Jonkisz 1990, **D22**, pp. 242-246).

Our measure of significance is a numerical function. The number standing for the importance (size, magnitude) of the transition m is identical to the cardinality of the symmetric difference of maximal expositions of nets of a given transition. The measure defined in such a way can be determined for any transition, because transitions, no matter whether within one theory or leading to a new theory, are always represented by pairs of nets. It is important that this way of establishing the size of transitions ensures commensurability of both of these kinds of transitions.

Using maximal expositions is, as we know, necessary to expound the whole content of a theory, which makes it possible to grasp similarities and differences between the original and final states of any transition. If, however, establishing the size of a transition required maximal expositions of both nets, then determining the value of F_m^{mx} would be practically unattainable. Employing symmetric difference ensures effectiveness, because it allows one to terminate expositions of both nets (versions of theories) on reaching the strongest, i.e. the most detailed, common assumptions. Determining the limit of exposition, namely finding the core K^*, the scope I^*, or the element T^* limiting the exposition (in the sense of **D4.B**) depends on the nets (versions of theories) making up the transition. Common assumptions are found much more easily if one compares two versions of one theory: those assumptions (common laws of various character, or postulates concerning their scope of application) are in such a case numerous; they can be seen in many elements of both nets including their basic elements. Common assumptions of different theories are less visible; they are more remote. Finding them requires going beyond the theories compared, but usually not beyond physics. Such assumptions come from the most fundamental physical theories. They can be principles common to many physical theories; at any rate, one does not have to look among them for a statement as general as the thesis that something exists (the only way of testing the supposition that the search will not lead beyond physics is actually comparing particular theories). Concentrating on the strongest common assumptions is justified because theory-elements whose cores are more general either do not figure in the determined differences of maximal expositions (if they belong to both expositions) or – if they do appear because their identical cores are in both nets connected with different scopes – their number can easily be estimated on the basis of cardinalities of extensions assigned to them in the theories under comparison[20].

---

20 In (Jonkisz 1990) one can find formulae facilitating determining dimensions of a transition, among them ones concerning cardinality of the set of expositions of a given element limited by a given core (*ibid*, **Ld20a-c**, pp. 217-218; 418-419); it has also

Soundness of the universal measure of significance is confirmed by many theorems. It is possible to show that:

(i) values of the significance function are greater the greater are differences between the original and the final states of a transition;

(ii) changes have no significance if and only if a transition is an identity (is stagnation);

(iii) any two symmetric transitions $\langle N, N' \rangle$, $\langle N', N \rangle$ have the same significance;

(iv) significance of the transition $\langle N_p, N_k \rangle$ is at most equal to (it is usually smaller than) the sum of significances of any intermediate stages leading from N_p to N_k (e.g., $\langle N_p, N_1 \rangle$, $\langle N_1, N_2 \rangle$, $\langle N_2, N_k \rangle$).

Application of this theorem to any activity yields the following practical rule: it is better to carry out changes at once rather than gradually, because their cost is never greater than the cost of gradual transition and the achieved goal is the same.

The properties (ii–iv) show that F_m^{mx} is a metric function[21]. Therefore, calculating the value of this function is a general, effective, and appropriate way of establishing the size of a transition, i.e. the distance between any scientific results: any versions of any theory.

5.2 The numerical characteristic of progress

Properties of the function F_m^{mx} justify employing it in a quantitative characterization of progress. In this application, it is also an appropriate and natural measure: the measure of progress.

(D6) If $m = \langle N, N' \rangle$ is a transition in the development of a discipline, m_p is a progressive and m_r a regressive aspect of the transition m (with respect to a particular net of results), then:

A. the degree of progressiveness of the transition m, i.e. p_{m_p}, equals $F_{m_p}^{mx}$;
the degree of regressiveness of the transition m, i.e. p_{m_r}, equals $F_{m_r}^{mx}$;
the degree of progress achieved as a result of the transition m, i.e. p_m, equals $p_{m_p} - p_{m_r}$.

B. The transition m is:
globally progressive *iff* $p_m > 0$;

been shown that adopting the more convenient way of determining significance does not influence the correctness of achieved results (*ibid*, **Ld22**, pp. 246-247; 430-431).

[21] Cf. (Jonkisz 1990, **T35-T38**, **W6**, pp. 246-251; 431-434).

globally regressive *iff* $p_m < 0$;
neutral with respect to progress *iff* $p_m = 0$.

The concepts of progressive and regressive aspects of transitions make it possible to locate parts of the transition *m* corresponding to progress or to regress, respectively. The concept of significance creates the possibility to evaluate the size of both of the aspects; therefore, the degree of progressiveness is identical with the value of the function F^{mx} for the progressive aspect m_p of the transition *m*, while the degree of regressiveness is identical with the value of the function for m_r. The degree of progrèss p_m stands for the balance of changes of both kinds and therefore it is calculated as a difference between the degrees of progressiveness and of regressiveness. This definition also covers cases of (global) regress, which are specified as transitions with the negative value of progress p_m. If a particular transition *m* cannot, on the basis of those criteria, be declared as either progressive or regressive, then *m* is a transition neutral with respect to progress. It should be clear that within a transition that is neutral according to our definition, it is usually possible to single out particular progressive modifications. However, the totality of such transitions (i.e. their general significance or magnitude) are neutralized by equally significant regressive changes.

It can be shown that the above way of relating the universal measure of significance to the concepts of progressive and regressive aspects of transitions is sound. Our way of establishing degrees of progressiveness, regressiveness and progress of any transition *m* is in full agreement with intuitions contained in the concepts of full progressiveness and full regressiveness of the transition *m* (defined on the basis of **D3**). Self-evident requirements related to these concepts are verified:

(i) every entirely progressive transition *m* (i.e. such that $m_{rg} = \langle\varnothing, \varnothing\rangle$) should have a zero level of regressiveness and *vice versa*;
(ii) every entirely progressive transition is generally progressive and every entirely regressive one is generally regressive (it has a negative degree of progress) unless it is stagnation;
(iii) entirely progressive and entirely regressive transitions can differ with respect to the degree of progress[22].

The last conclusion clearly demonstrates the difference between qualitatively characterized full progressiveness (and full regressiveness) and quantitatively characterized degree of progress. A proclamation of full pro-

[22] Cf. (Jonkisz 1990, **W14a-d**, pp. 305-306; 451-453). It is also easy to conclude that the relation ordering transitions with respect to their progressiveness provides good ordering of all transitions (*ibid*, **W15**, pp. 306; 453).

gressiveness of a transition means that the transition has no regressive side. The degree of progress is an outcome of the importance of progressive and regressive changes, so it is possible that it has different values even within transitions having no regressive side: such transitions can differ in the size of their progressive aspects, which exhaust given transitions, and can – unless they are cases of stagnation – differ in their degree of progress.

The way of establishing degrees of progressiveness and regressiveness is also in agreement with results obtained on the basis of the universal measure of significance. The progressive and regressive aspects exhaust any transition; therefore, it is justified to expect that the sum of the two degrees should be equal to the general size of a transition obtained by means of the universal function of significance: $F^{mx}_{m_p} + F^{mx}_{m_r} = F^{mx}_{m}$[23].

The concepts enabling a quantitative characteristic of progress work also in specially selected, diagnostic types of transitions.

(1) Cases of stagnation ($m = \langle N, N \rangle$) are usually labelled without hesitation as neither progressive nor regressive, and if one requires progress to exhibit at least minimally beneficial changes, then stagnation is regarded as regressive. In the conception of relative progress, no judgement can ever be so unequivocal. Maintaining some state of science is never neutral to relative progress. Maintaining the correct components is progressive while maintaining the incorrect ones is regressive. The distinction between the correct and the incorrect components depends on the choice of the net of results N_w; therefore, it is obvious that the same state of stagnation can be judged as entirely progressive, partially progressive, partially regressive or even as entirely regressive, depending on the accepted net of results. However, evaluation of cases of stagnation with respect to their degree of progress is unequivocal. The degree of progress shows how much a given transition has pushed the development of science toward the state N_w. The degree of progress understood in such a way must in the case of stagnation equal zero, regardless of the choice of the net of results: simply because stagnation does not move us toward any end. The set of concepts used in the conception of relative progress makes it possible to confirm all these results[24].

(2) Transitions terminating in a given net of results N_w: $m = \langle N, N_w \rangle$. Their peculiarity consists in that the final point of the transition m is at the same time the basis of evaluation of progress achieved as a result of m. It is a historical rule that authors of such changes always judge them

23 Cf. (Jonkisz 1990, **T42**, pp. 306-307; 454).

24 Cf. (Jonkisz 1990, **W16a-b**, pp. 307-308; 455-456).

to be progressive. This regularity is confirmed by our conception; one can confirm that each such transition is entirely progressive.

(3) Growth and progress, decline and regress. If the original net of the transition $m = \langle N, N' \rangle$ is embedded in its final net ($N \, \mathrm{p}_{\mathrm{in}} \, N'$), where $N \neq N'$, then the transition exhibits growth; the reverse direction of the relation p_{in} is a sign of decrease or decline. If one does not accept the conception of relative progress, i.e. if one evaluates only contextual progressiveness, then progress is thought to be identical with growth, and regress – to be identical with decline. On the basis of our conception, we obtain a different solution: cases of growth intersect with progressive transitions, and cases of decline intersect with regressive transitions. There is, then, both progressive and regressive growth, as well as progressive and regressive decline. Cases of entirely regressive growth and entirely progressive decline are also possible.[25]

(4) Bifurcation of progress. In such cases, a certain state of knowledge is developed in two different, conflicting directions. Those cases can be represented in our terms by a pair of transitions m_1 and m_2 such that $m_1 = \langle N, N_1 \rangle$, $m_2 = \langle N, N_2 \rangle$ and neither $N_1 \rho N_2$ nor $N_2 \rho N_1$. There is a valid question whether both such transitions can be progressive with respect to one and the same net of results. This question can be answered; it is possible to show that both such alternative transitions can be progressive with respect to a particular net N_w (even to equal degrees), although neither $\langle N_1, N_2 \rangle$ nor $\langle N_2, N_1 \rangle$ are progressive with respect to that net[26].

(5) The degree of progress of final transitions, i.e. of ones terminating in the net of results. The first question that might arise in connection with such transitions is: What is the relation between the degree of progress of the final transition $m = \langle N, N_w \rangle$ and degrees of progress of joined transitions beginning in N and terminating in N_w? It can be shown that the degree of progress of the transition $m = \langle N, N_w \rangle$ equals the sum of degrees of progress of transitions making up the sequence from N to N_w[27].

Thanks to the above regularity, it can be proved that the degree of progress of the final transition $m = \langle N, N_w \rangle$ is greater than the degree of progress associated with any other transition, no matter how long, beginning in N and not terminating in N_w. This is so, because if the transition $m = \langle N, N' \rangle$ does not terminate in N_w, then another transition, linking N' with N_w, is required. The degree of progress of the transition

---

25 Cf. (Jonkisz 1990, **T44**, pp. 309; 456-457).

26 Cf. (Jonkisz 1990, **T45**, pp. 310-311; 457-458).

27 Cf. (Jonkisz 1990, **T47**, pp. 315; 460-462).

$\langle N, N_w \rangle$ equals the sum of degrees of progress of the transitions $m' = \langle N, N' \rangle$ and $\langle N', N_w \rangle$, therefore $p_m > p_{m'}$.[28]

6. Effectiveness of progress

To evaluate any changes with respect to progress, it is essential to take into account not only the final result but also the way in which progress was achieved. This aspect of progress can be grasped by means of the concept of effectiveness.

(D7) Effectiveness e_m of the transition m is equal to the quotient p_m / F_m^{mx}; Effectiveness E_{m_i} of the sequence $\{m_i\}$, $i = 0, 1, ..., k$, of joined transitions is equal to the quotient of the sum of degrees of progress and the sum of significances of particular transitions from $\{m_i\}$: $E_{m_i} = \sum_{i=0}^{k} (p_{m_i}) / \sum_{i=0}^{k} (F_{m_i}^{mx})$.

The concept of effectiveness enables us to take into consideration the cost of achieving progress. The effectiveness is the ratio of achieved progress to the size of a particular transition. The effectiveness of a sequence of joined transitions (in which a subsequent stage is a continuation of the prior one) is defined as the ratio of the sum of stages of progress to the sum of significances of particular component transitions. After combining the concept of progress with our results concerning degrees of progress and of significance, we obtain the formula for the coefficient of effectiveness:

(*) $\quad e_m = (p_{m_p} - p_{m_r}) / (p_{m_p} + p_{m_r})$.

Progress can be easily expressed in terms of effectiveness and length, i.e. fundamentality of a transition: progress depends on a combination of effectiveness and significance ($p_m = e_m F_m^{mx}$)[29].

An analysis of the quotient describing the way of determining the degree of effectiveness leads us to conclusions compatible with our intuitions regarding progress and effectiveness. The denominator of the formula for effectiveness is always non-negative, i.e. it equals zero only if a transition is stagnation, and it is positive in all other cases. Therefore, the value of the effectiveness coefficient depends on the sign of the numerator, which is the same as the sign of the degree of progress

[28] Cf. (Jonkisz 1990, **T48a**, pp. 316; 462).

[29] Cf. (Jonkisz 1990, **D32**, **W18**, pp. 316-318; 463).

of the transition m. Thus, in the case of general progress, the effectiveness coefficient is positive, and in the case of general regress the coefficient is negative. Effectiveness reaches the boundary value of zero in the case of transitions neutral with respect to progress, except for a special kind of a neutral transition, i.e. stagnation, in which the denominator of the expression (*) also equals zero. Effectiveness is thus undetermined only in the case of stagnation, which is also in agreement with our intuitions: there is no effectiveness to be discussed if nothing has been achieved ($p_m = 0$, neutrality with respect to progress) and if no cost has been borne ($F_m^{mx} = 0$).

Additionally, an analysis of the formula (*) enables us also to determine the maximum ($e_m = 1$) and the minimum ($e_m = -1$) values of the effectiveness coefficient. Also, it is easy to conclude that the maximum value characterizes entirely progressive transitions ($p_{m_r} = 0$), while any entirely regressive transition is minimally effective ($p_{m_p} = 0$). Therefore, all final transitions $\langle N, N'_w \rangle$ are maximally effective (because they are entirely progressive), regardless of what point they originate from. Greater costs, necessary if the starting point is more distant from N_w, are counterbalanced by the size of achieved progress, if no additional cost has been borne, which is the case with transitions leading straight to the goal[30].

Those conclusions agree with the concept of (progressive and regressive) aspects of transitions. As the regressive aspect of the transition m grows, general progress gradually decreases. Once the critical value of zero (neutral transition) is passed over, we face general regress. A negative value of the effectiveness coefficient in the equation $p_m = e_m F_m^{mx}$ shows that the transition m is moving away from N_w. The absolute value of the coefficient indicates the direction of movement; when effectiveness reaches the boundary value of minus one, it means that changes of the transition m are going in the direction diametrically opposed to the right one. The minimum value of the effectiveness coefficient ought not to be confused with the value p_m of the degree of general regress. This degree has no minimum value. Its absolute value can grow indefinitely because there is no maximum length (F_m^{mx}) for the transition leading away from N_w. In short, there are no restrictions on mistaken solutions.

7. Direction of changes and progress

The conception of relative progress makes it possible to give a more detailed account of the direction of changes thanks to the geometrical in-

---

30 Cf. (Jonkisz 1990, **T49**, **W19**, pp. 318-320; 463).

terpretation of concepts of the present approach. We take changes to be transitions between accepted results, namely as transitions between particular versions of a theory or theories. We treat theories as points in the space $\boldsymbol{N}$ (i.e. in the class of all theory-nets), and it is the universal function of significance that determines distance in this space. Changes (transitions) are thus represented as vectors. Absolute values of vectors defined in this way are identical to the distance F_m^{mx} between any two nets N and N' defining transitions $m = \langle N, N' \rangle$. Their senses are the same as senses of the transitions, i.e. from N to N'.

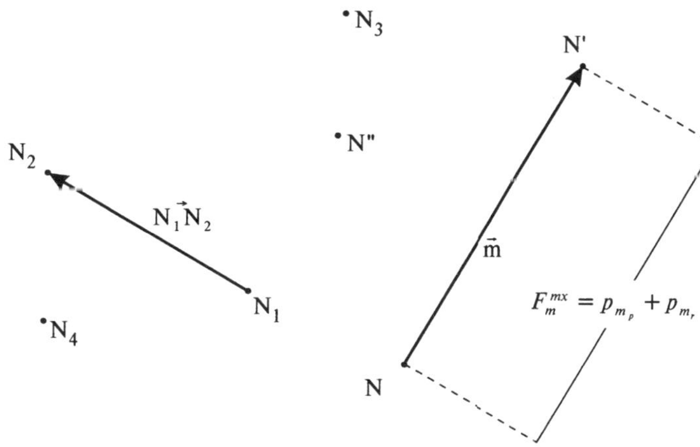

Figure 3

In order to identify directions of vectors representing changes, it is first necessary to orient the space in a way compatible with the conception of relative progress. Therefore, we single out from among all points of the space $\boldsymbol{N}$ the point N_w representing the net of results and define the preferred direction by means of the vector $\vec{NN_w}$. This vector is called the directional vector; the line on which it is located is called the directional axis. The direction of the vector $\vec{m}$ $(m = \langle N, N' \rangle)$ is the angle between $\vec{NN'}$ and the directional axis (called the angle of deviation).

We need a method for determining the direction – understood in the above way – for every transition and for comparing directions of any two transitions in the space $\boldsymbol{N}$. This method consists in placing vectors characterizing particular transitions. The method can be summarized as follows: by means of translation, convert vectors of the transitions into co-planar vectors with a common origin; and even more precisely:

vectors corresponding to transitions under comparison ought to lie in the same half of the plane (let us assume in the right half)[31].

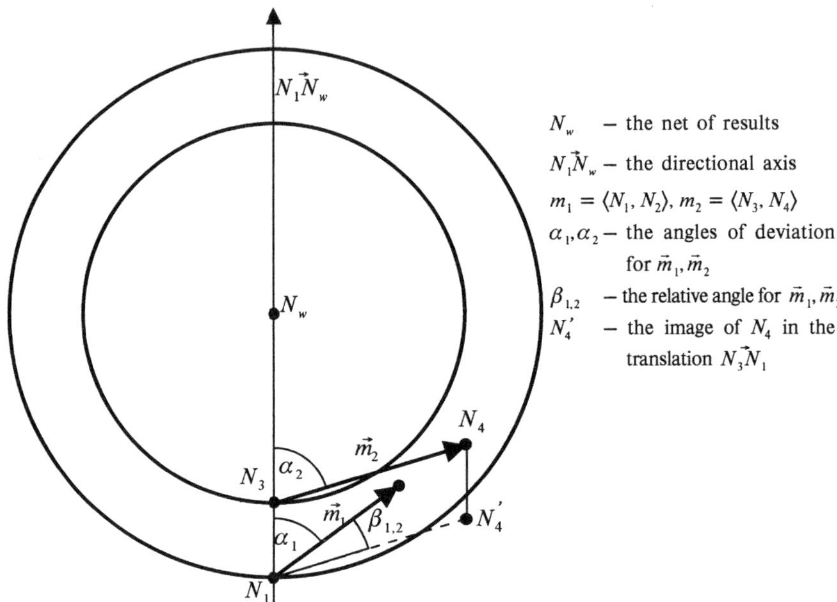

Figure 4

On the basis of what has been said up to this point, we can define the following: the angle of deviation of any transition *m* (determined by the vectors $\vec{m}$ and $\vec{NN}_w$), the relative angle for any pair of transitions *m* and *m'* (determined on the basis of the vectors $\vec{m}$ and $\vec{m}'$), and the so-called forking angle (i.e. the relative angle of transitions with the same origin)[32].

In order to determine the relations between the direction of changes and progress, it is necessary to develop our geometric interpretation further, so that we can relate the definitions of angles of deviation and of relative angles to our results concerning significance (length) of changes and to results concerning progress. Our goal is to determine the relation between the value p_m of general progress and its compo-

[31] It can be shown that this method makes it possible to compare directions of any transitions. There always exists a plane transformation ensuring such comparability: it is a composition of at most three rotations and at most one translation. Cf. (Jonkisz 1990, pp. 323-324).

[32] Cf. (Jonkisz 1990, **D33**, pp. 324-326).

nents p_{m_p}, p_{m_r} and the vector $\vec{m}$ of any transition $m = \langle N, N' \rangle$. Discovering those relations will let us determine angles of deviation and relative angles just by means of numbers characterizing progress.

General progress p_m and its components p_{m_p}, p_{m_r} can be represented as vectors. Since p_m signifies progress achieved as a result of the transition $m = \langle N, N' \rangle$ (progress with respect to the net N_w), it will appear in our geometrical interpretation as the vector $\vec{p}_m$ originating from the point N with the same direction as the vector $\overrightarrow{NN}_w$. The module of the vector $\vec{p}_m$ equals the degree p_m obtained according to **D6.A**, while its sense depends on the sign of the degree of progress. If the sign is positive, then the sense agrees with that of the vector $\overrightarrow{NN}_w$, and if the sign is negative, then the sense is opposite to it. In the space N, we can also distinguish components of the vector $\vec{p}_m$, namely the vector $\vec{p}_{m_p}$ and $\vec{p}_{m_r}$ (**D6.A**). According to our understanding of degrees of progressiveness and regressiveness, we place the origin of both of these vectors at point N. Next, we give the vector $\vec{p}_{m_p}$ the same sense as the vector $\overrightarrow{NN}_w$ has, and the vector $\dot{p}_{m_r}$ – the opposite one. Since simultaneously $p_m = p_{m_p} - p_{m_r}$, we can see that the vector $\vec{p}_m$ is the (vector) sum of the two components with opposite senses.

The general picture is then the following. The vector p_m of progress indicates approaching (when $p_m > 0$) the net of results N_w or moving away from it (when $p_m < 0$), due to the transition m. Components of the vector $\vec{p}_m$, i.e. $\vec{p}_{m_p}$ and $\vec{p}_{m_r}$ give an account of magnitudes of the progressive and the regressive sides, or inform us what general progress would be if there were no opposite side. The length of the transition m, i.e. F_m^{mx} is, as we know, the sum of values of these components of general progress $F_m^{mx} = p_{m_p} + p_{m_r}$.

Concepts describing the direction of changes and concepts characterizing progress are connected by the following relation:

(**) The distance of the final state of any transition $m = \langle N, N' \rangle$ from the net of results N_w should correspond to the general progress of the transition.

The end of the vector p_m, which we shall call N_p, should then be at the same distance from $N_{w'}$ as the end N' of the vector m. Therefore, both of these points will be located on the same circle, whose centre lies in N_w. Let $R_{N'}$ stand for the radius of the circle, and $oR_{N'}$ – for the circle itself. Adopting these symbols makes it obvious that $R_{N'} = R_{Np}$ and that $oR_{N'} = oR_{Np}$.

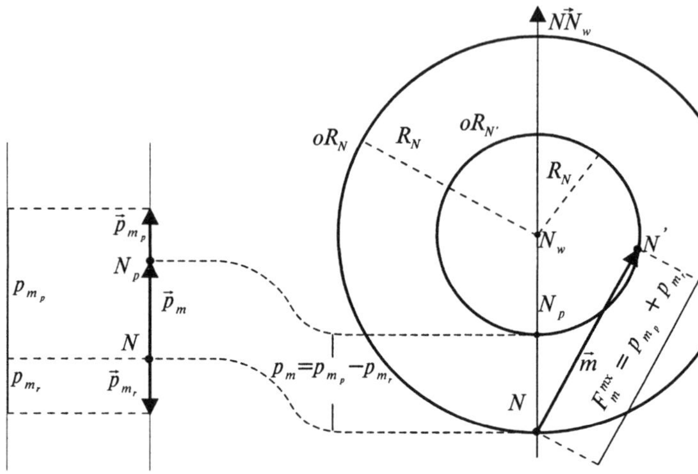

Figure 5

The relation (**) opens the way to providing a numerical characteristic of angles of deviation and relative angles. Equations determining the magnitudes of those angles contain only parameters describing progress[33]. Such a characteristic makes it possible to relate concepts basic to the conception of relative progress to known concepts and results in geometry. It is possible to express in geometric terms, e.g. the concept of effectiveness of a transition[34].

The so-called class of instances of equal progress interpreted geometrically appears to be a useful concept. It is understood as the set of all transitions whose original points are at the same distance from the net N_w and which achieve the same general progress (which can be positive, negative or equal to zero). To determine this class, it is necessary only to possess information about the radius of the circle of original points and about the radius of the circle of final points – because the centre of both of the circles is the point N_w. For this reason, the class of instances of equal progress shall be represented as $\{R_N, R_{Np}\}$, where R_N is the radius of the circle of original points, and N_p is the final point of the vector of general progress, characteristic of all transitions of the class. Consequently, R_{N_p} is the distance of final points of the transitions from the net N_w.

33 Cf. (Jonkisz 1990, **Ld33a**, pp. 329-330; 464-466).

34 Cf. (Jonkisz 1990, **Ld33b**, pp. 330; 466).

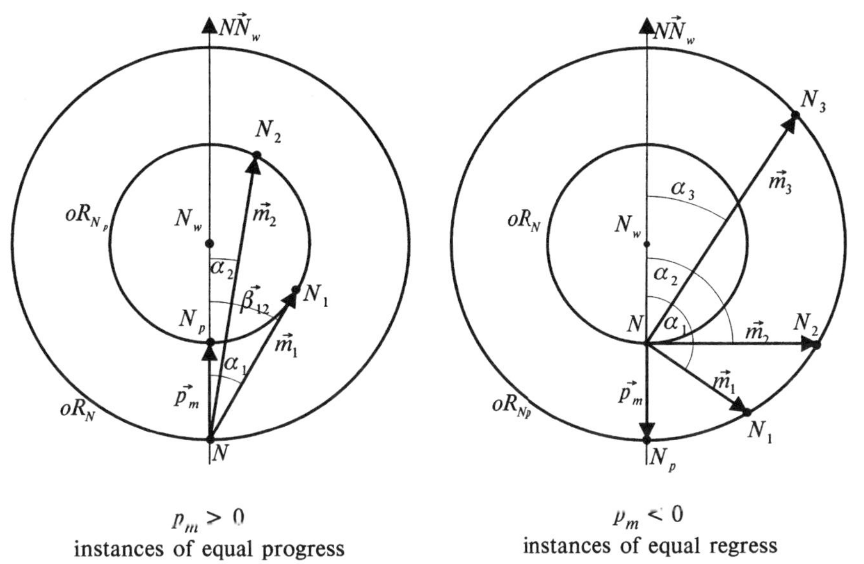

Figure 6

The geometrical interpretation of the class of instances of equal progress makes it possible to justify the conclusion that there is a relatively very small number of transitions whose length equals the difference of both radii, namely transitions satisfying the condition $F_m = |p_m|$. It is easy to confirm that in such cases we are dealing with a very rare kind of transition, i.e. with entirely progressive or entirely regressive changes. It suffices to compare the equations $F_m = p_{m_p} + p_{m_r}$ and $p_m = p_{m_p} + p_{m_r}$ with the condition describing such transitions to reach the conclusion that the condition can be satisfied only in two cases: if $p_{m_r} = 0$ or if $p_{m_p} = 0$ – that is, if the transition is either entirely progressive or entirely regressive. Under the geometrical interpretation, each of such shortest transitions $m = \langle N, N_p \rangle$ to the end-circle corresponds to infinitely many longer transitions, that is to the totality of transitions originating in N and ending in any point of the end-circle, except for the point N_p.

The geometrical approach facilitates the noticing of the difference between instances of stagnation and transitions neutral with respect to progress. A given set of neutral transitions, that is a special kind of instances of equal (zero) progress, is represented by $\{R_N, R_N\}$ – so it is represented by one circle R_N, which is simultaneously the set of original and final points of all transitions of the class. Instances of stagnation (at a given distance from N_w) are neutral transitions with zero length, while the majority of transitions neutral with respect to progress have nonzero length. Every instance of stagnation $\langle N, N\rangle$ corresponds to an infinite set of nonidentity-type neutral transitions, i.e. to the totality of transitions originating in N and ending in $N' \neq N$, which lie somewhere on the circle oR_N.

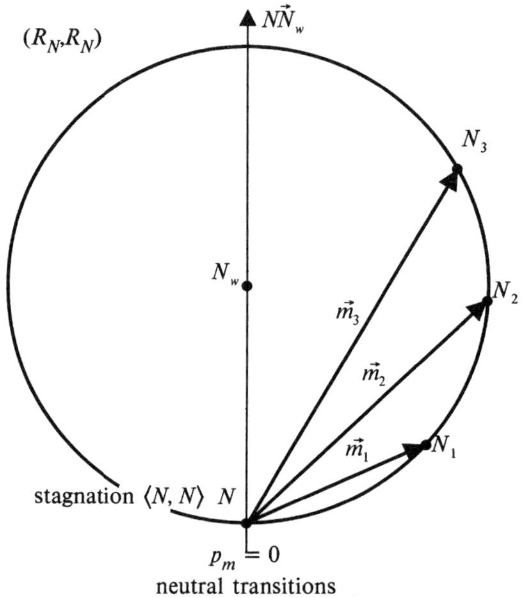

Figure 7

The geometrical picture also makes it easier to determine minimal and maximal length of a transition in the class $[R_N, R_{Np}]$ of instances of equal progress. We can notice that the minimal length of transitions in this class equals $|R_N - R_{Np}|$ and that the maximal length is given by the formula $2R_N - (R_N - R_{Np})$. These equations are satisfied regardless of whether the class is the class of global progress, global regress (i.e. negative degree of progress), or the singular class of neutral transitions[35].

The geometrical approach makes it possible to determine values that angles of deviation of a certain type can take (let us bear in mind that we mark such angles only in the right half-plane). The degree of

[35] The formula for the maximal length given here corrects a mistake in (Jonkisz 1990, **Lt51a**, p. 332).

deviation of entirely progressive transitions equals zero, and that of fully regressive ones – 180 degrees. The globally progressive$_{\rightarrow}$**(D6.B)** transition $m = \langle N, N' \rangle$ can deflect from the directional vector NN_w from zero to 90 degrees but it cannot assume the boundary value of 90 degrees. Neutral transitions have the same range of values, but the angle of deviation of identity-type neutral transitions (i.e. stagnations) is undetermined. However, the range of values is broadest for globally regressive transitions and stretches from zero to 180 degrees[36].

It is interesting that the range of regress partially overlaps (in this respect) the range of global progress. Moreover, among globally regressive transitions there can be ones with a zero degree of deviation, which means that they are directed toward the net of results N_w. Let us divide all transitions directed toward N_w into globally progressive and globally regressive and let us first examine instances of global progress. A globally progressive transition is any transition whose progressive side exceeds its regressive side (in the sense defined in **D3**). A separate group in the class of instances of global progress is made up of entirely progressive transitions, that is, of transitions without a regressive side. Among them, we can distinguish maximally progressive transitions, i.e. ones realizing maximal progress that can be achieved with N as a starting point. It is obvious that all transitions ending in N_w have this property. All other transitions, including all transitions directed toward N_w, have smaller degrees of progress. That can result from the fact that such transitions either do not reach N_w (in this case they are entirely but not maximally progressive) or they go beyond that point (and cease to be entirely progressive). Regressive transitions directed toward N_w are simply transitions going beyond the point N_w[37].

The geometrical approach also makes it possible to define easily regularities which determine whether a transition directed toward the net of results is entirely progressive or if it is progressive at all. That is decided by the length of the transition $m = \langle N, N' \rangle$. If the length of F_m is greater than R_N (the distance of the point N from N_w), then the transition ceases to be entirely progressive. If the length is greater than $2R_N$, then the transition m is an globally regressive transition, in spite of being directed toward the net N_w[38].

Those conclusions agree with the fundamental idea of relative progress: progressive is what has brought development of science closer to results regarded as correct. Approaching the state N_w from the point N is purely progressive on the segment NN_w but exceeding the point N_w and

36 Cf. (Jonkisz 1990, **T51a-f**, pp. 333-336; 466-469).

37 Cf. (Jonkisz 1990, **W20**, pp. 336; 469).

38 Cf. (Jonkisz 1990, **W21a-b**, pp.336-338; 470-471).

going further toward the final point N' ceases to be correct (from the perspective of N_w). We can clearly see that in the light of a new, entirely progressive transition $\langle N', N_w \rangle$. Losses sustained on the segment $N_w N'$ count as a regressive part of the transition m and if transgression beyond N_w is too large, then the transition eventually leads to regress. Dependence of progress on the direction and size of changes has been unveiled again. Even changes proceeding in the right direction cease to be progressive if they go too far.

8. Evaluation of the conception of relative progress

Let us begin our evaluation of the conception of relative progress by examining its structure. The conception rests on a new understanding of progress as an objective but relative property of scientific development or, rather, of changes in science constituting its development. Such an understanding of progress is independent of the structuralist conception of theory and of methods of the structuralist view. However, most of the conception based on such an understanding of progress could not be developed without applying methods of that view. There seems no way of attaining numerical and geometric descriptions of changes and progress without applying the structuralist method of reconstructing theories[39].

The concept of reduction employed in our conception of progress is not essential to it, in the sense that the whole conception will not change if we replace our definition of reduction with a better one. The concept of reduction is an easily interchangeable component of our conception.

The numerical characteristic of progress is based on the concept of fundamentality of a transition (the function F_m), which is independent of the relative understanding of progress. This function is appropriate and, at the same time, it has metric properties. Therefore, it can be treated as a measure of distance in the space N, whose points are theory-nets (but only a minute part of the whole space of theory-nets is

39 The structuralist theory reconstruction – and *a fortiori* the whole view – should not be reduced to the method of axiomatization of theories introduced by P. Suppes and called axiomatization by definition of a set-theoretic predicate. Since J. Sneed applied that method more widely, much has been achieved in terms of both methods and the structuralist conception. The most important of the later achievements is a new way of reconstructing the so-called special laws, which makes it possible to present a theory as a set of theory-elements ordered by the relation of specialization; cf. (Balzer *et al.* 1977/78). Without such an approach to the structure of physical theories, it would not even be possible to pose many of the problems solved in the conception of relative progress.

occupied by points corresponding to versions of theories proposed in the history of science). Changes in science are represented as transitions $m = \langle N, N' \rangle$ from one net to another, that is as vectors. Precise analyses of progress have been made possible by providing a way of orienting the space *N*; the concept of the net of results finds its place there. Only after taking the point N_w as the central point of the space *N* can one begin to consider a way of judging whether and to what degree particular transitions move us closer to the goal – to N_w.

It is necessary to explain why we called the net of results the goal. The context excludes the meaning of the word "goal" used in talking about consciously working toward a goal. Such conscious effort is possible only if the goal is known and attainable. A necessary condition of attainability of a goal is, in turn, its being real.

The net of results is surely real, in the sense that it is a collection of results achieved in science. Let us sidestep debates about whether a collection has the same reality as its elements and what exactly in science corresponds to a net of results established *ex post* as a result of metascientific reconstructions. We juxtapose reality of the net of results with ideal goals cherished possibly by some scientists (e.g. maximal verisimilitude), which are surely outside the realm of actual science.

The very definition of the net of results implies that it is an attainable goal, and that it cannot be a goal known in advance. So, it cannot be the case that scientific theories are the results of proceeding toward a net of results. It is quite the opposite; it is the theories we attained and accepted that form the net N_r.

The most important feature of our goal is that the net of results changes. Thus the vision of one goal of all scientific work – the goal with respect to which one could proclaim progress – is discarded.

Is it necessary to associate the following: a real but changeable goal with the conception of relative progress and a constant but ideal goal with the conception of absolute progress? It would be better if it were not, because we could then build a conception of progress offering us working criteria of absolute progress. However, the development of the general theory of scientific progress seems to confirm permanence of the association. Acceptance of absolute progress is related to an idealistic conception of the goal of science and it leads to ineffective criteria of progress[40].

---

40 The conception of progress presented in this volume by C. Moulines also falls under this category. The criteria of progress he suggests to apply in cases of partial incommensurability of theories can be summarized as follows. The set of appropriately described domains belonging to the non-empty set common to both theories grows (the

Opinions formulated in the present paper do not bar a search for working criteria of absolute progress, which should rather be called progress with respect to a constant, ideal goal of science. The choice of the conception of relative progress is justified not only by the fact that it enables a formulation of working – even numerical – criteria of progress, but also by the fact that assumptions of idealistic conceptions conflict with the general picture of science we accepted at the beginning of this paper (Fig. 2). Teleologico-idealistic conceptions of progress can only fit the vision of scientific progress depicted in Fig. 1: the goal would then be to cover fully the realm of ignorance with all-embracing knowledge. But if ignorance grows together with knowledge, then the very definition of the ideal goal (e.g. that science should progress toward an all-embracing theory) leads, under this assumption, to a contradiction: it

---

conditions *a* and *b* of the definition *PI*). The number of appropriately described empirical domains specific to the later theory is greater than the number of such domains of the theory being eliminated (the condition *c*) and the number of improperly described specific domains decreases on the way from *N* to *N'* (the condition *d*). Still, in spite of Moulines' arguments, the assumption of existence of a non-empty overlapping part of ranges of both theories is doubtful. If one assumes the possibility of describing phenomena independently of terms of both theories, then one cannot refer to the overlap part of the ranges (in the sense of $I \subseteq M_{pp}$), but only to some "plausible correlation" of domains of both ranges. The latter way of comparing ranges clearly leans towards macrological reduction and it suggests, or maybe even straightforwardly assumes, that such correlations are rendered from a later point of view, from which one can clearly see both juxtaposed theories. Such suggestions gravitate towards our net of results. The conditions *b*-*d* also raise objections. If one judged them as merely defining the concept of progress, without pretending to provide criteria, then one could notice they assume an ideal goal of science. Required growth in the number of appropriately described empirical domains (in the sense of domains belonging to *r*(*M*), that is such in which laws of the theories are satisfied) and a decrease in the number of improperly described ones (laws of the theories are not satisfied in them) assumes that the ideal goal is a true and all-embracing theory. Relating empirical domains to models of a given theory also makes it difficult to test the conditions, which undermines their status as criteria. The explanation that it is possible to determine cardinality of ranges of theories (cf. note 14) raises no protest, provided one understands ranges – as Moulines explains – as applications postulated by scientists and – this is worth noting and is important to our criteria of progress as well – if one wants to estimate not the number of particular applications but the number of classes of applications representing phenomena in the general sense, not their particular instances. This explanation is not sufficient, however, because one would count domains described either appropriately or improperly. That would call for an effective conception of empirical confirmability, and is prone to the same objections as were put forward against definitions of progress based on the concept of verisimilitude.

follows from the definition of the goal that there is an area of ignorance independent of knowledge that can be engulfed by growing knowledge.

Summing up, relativity of our conception of progress has no dangerous consequences usually associated with the claim of the relativity of progress. Namely, our conception does not lead to epistemological relativism as progress is not related to the concept of truth; therefore, it does not follow from the claim that a theory is at one time progressive and at another time it is not, that it is at one time true and at another time false[41]. Progressiveness is a relative but objective property. Once one establishes a basis of judging progress, then for any change it is possible to justify only one of the two possible claims regarding progress.

Institute of Social Sciences
Silesian University in Katowice
Cieszyn Branch
Bielska 62
43-400 Cieszyn, Poland
and
Department of Social Sciences
Technical University of Łódź
Bielsko-Biała Branch
Willowa 2
43-300 Bielsko-Biała, Poland

REFERENCES

Balzer, W., Moulines, C.U. and Sneed, J.D. (1987). *An Architectonic for Science.* London: Reidel.

Balzer, W. and Sneed, J.D. (1977/1978). Generalized Net Structure of Empirical Theories. *Studia Logica* **36**, 195-221 and **37**, 167-194.

Hajduk, Z. (1989/1990 and 1991/1992). Uwarunkowania postępu poznawczego w teoriach postępu nauki. Część I i II [Criteria of Epistemic Progress in Various Theo-

[41] The issue presents itself differently in truth-based conceptions of progress and in all teleologico-idealistic conceptions, as well. They assume an ideal, constant goal of science, e.g. certainty – in the sense of probability equal one – of theory's statements or an ideal of all-embracing true knowledge. If one maintained within such a conception that a theory is and, at the same time, is not progressive, then it would mean that it is approaching the goal and, at the same time, it is not, e.g. it is and, at the same time, it is not closer to truth, or simply – that it is and it is not true. Relativism of truth would then be evident.

ries of Scientific Progress. Part I, and Part II]. *Roczniki Filozoficzne* **XXXVII/XXXVIII**, z. 3, 83-160; and **XXXIX/XL**, z. 3, 23-56.

Jonkisz, A. (1984). Zagadnienie postępu w ujęciu strukturalistycznym [The Problem of Scientific Progress in the Structuralist View]. *Roczniki Filozoficzne* **XXXII**, z.3, 149-172.

Jonkisz, A. (1990). *Struktura, zmienność i postęp nauki. Ujęcie strukturalne.* [The Structure, Change, and Progress of Science. The Structuralist View]. Lublin: Wydawnictwo UMCS.

Lakatos, I. (1970). Falsification and the Methodology of Scientific Research Programmes. In: I. Lakatos and A. Musgrave (Eds.), *Criticism and the Growth of Knowledge.* Cambridge: Cambridge University Press, pp. 91-197.

Moulines, C.U. (1979). Theory-nets and the Evolution of Theories: the Example of Newtonian Mechanics. *Synthese* **41**, 417-437.

Sneed, J.D. (1971). *The Logical Structure of Mathematical Physics.* Dordrecht: Reidel.

Sneed, J.D. (1976). Philosophical Problems in the Empirical Science of Science: a Formal Approach. *Erkenntnis* **10**, 115-146.

Sneed, J.D. (1977). Describing Revolutionary Change: a Formal Approach. In: R. Butts and J. Hintikka (Eds.), *Historical and Philosophical Dimensions of Logic, Methodology and Philosophy of Science.* Dordrecht: Reidel, pp. 245-268.

Stegmüller, W. (1975). Structures and Dynamics of Theories. Some Reflections on J.D. Sneed and T.S. Kuhn. *Erkenntnis* **9**, 75-100.

Stegmüller, W. (1976). *The Structure and Dynamics of Theories.* New York: Springer.

Stegmüller, W. (1978). A Combined Approach to the Dynamics of Theories. *Theory and Decision* **9**, 39-75.

Turek, J. (1996) *Wszechświat dynamiczny. Rewolucja naukowa w kosmologii* [The Dynamic Universe. The Scientific Revolution in Cosmology]. Lublin: RW KUL.

CONTRIBUTORS

Wolfgang Balzer is professor of logics and philosophy of science at the University of Munich which position he has held since 1984. He has contributed to the philosophy of science (e.g. W. Balzer, C.U. Moulines, J.D. Sneed, *An Architectonic for Science. The Structuralist Program.* Dordrecht: Reidel, 1987) and to the theory of social institutions (e.g. A Basic Model of Social Institutions. *Journal of Mathematical Sociology* **17**, 1-29). He has provided a number of reconstructions of scientific theories, most recently from genetics (W. Balzer and C.M. Dawe, *Models for Genetics*. Frankfurt-Main: Peter Lang, 1997).

Piotr Giza (b. 1957) is a Associated Professor and director of the Computer Laboratory at the Faculty of Philosophy and Sociology of Maria Curie-Skłodowska University, Lublin, Poland. His research concentrates on the methodology and philosophy of Artificial Intelligence, in particular Automated Discovery Systems, as well as philosophy of science.

Adam Grobler (b. 1949) is professor of philosophy at the Tadeusz Kotarbiński Pedagogic University of Zielona Góra, Poland and Associated Professor at the Jagiellonian University, Kraków (part-time). His interests include methodology, theory of knowledge and philosophy of language, especially interdisciplinary problems of the above. His major publications: *Teza o niewspółmierności a problem redukcji teorii naukowych* (Incommensurability Thesis and the Problem of Reduction of Scientific Theories). Kraków-Wrocław: Ossolineum, 1986; *Prawda i racjonalność naukowa* (Truth and Scientific Rationality). Kraków: Inter Esse, 1993; *Względność i relatywizm* (Relativities and Relativism). Kraków: Aureus, forthcoming. He is the author of many translations of English-language philosophical literature into Polish, including selections of James', Hintikka's and Putnam's writings.

Kazimierz Jodkowski (b. 1950) is professor of philosophy and director of the Institute of Philosophy at the Tadeusz Kotarbiński Pedagogic University of Zielona Góra, Poland. His publications concern general philosophy of science and philosophy of natural sciences. Major of them include: *Teza o niewspółmierności w ujęciu Tomasa S. Kuhna i Paula K. Feyerabenda* (Incommensurability Thesis in Tomas S. Kuhn's and Paul K. Feyerabend's View). Lublin: Wydawnictwo UMCS, 1984; *Wspólnoty uczonych, paradygmaty i rewolucje naukowe.* (Scientific Communities, Paradigms and Scientific Revolutions). Lublin: Wydawnictwo

UMCS, 1990; *Metodologiczne aspekty kontrowersji ewolucjonizm-kreacjonizm* (Methodological Aspects of the Evolution/Creation Controversy). Lublin: Wydawnictwo UMCS, 1998.

Adam Jonkisz (b. 1953) is professor of philosophy and logic, vice-director of the Institute of Social Sciences and Cultural Studies, director of the Department of Philosophy at Silesian University in Katowice, Cieszyn Branch; and Associated Professor and director of Department of Social Sciences at the Technical University of Łódź, Bielsko-Biała Branch. His interests include general methodology of science and theory of knowledge, especially problems of scientific development and progress. His major publications include: *Struktura, zmienność i postęp nauki. Ujęcie strukturalne* (The Structure, Change and Progress of Science. The Structuralist View). Lublin: Wydawnictwo UMCS, 1990; *Ciągłość teoretycznych wytworów nauki. Ujęcie strukturalne.* (The Continuity of Theoretical Products of Science. The Structural View). Lublin: Wydawnictwo UMCS, 1998.

Leon Koj (b. 1929) is professor emeritus at the Catholic University of Lublin (part-time). His interests are: general methodology, logical semantics, an eventist's ontology, and logical foundations of ethics. His major publications include: *Semantyka i pragmatyka.* (Semantics and pragmatics). Warszawa: PWN, 1971; *Problemy semantyki logicznej* (The Problems of Logical Semantics). Warszawa: Biblioteka Myśli Semiotycznej, 1990; *Myśl i znak.* (Thought and Signs). Białystok: KAW, 1990.

Carlos U. Moulines (b. 1946) has been tenured professor at the National Autonomous University of Mexico, at the University of Bielefeld, and at the Free University of Berlin. Since 1993 he has been tenured professor and director of the Institute of Philosophy, Logic and Methodology of Science at the University of Munich. His principal fields of research are: general philosophy of science, history of science, axiomatization of physical and chemical theories, general ontological and epistemological problems, the work of Rudolf Carnap. His main publications are: W. Balzer, C.U. Moulines, J.D. Sneed, *An Architectonic for Science. The Structuralist Program.* Dordrecht: Reidel, 1987; *Pluralidad y Recursión.* Madrid: Alianza, 1991; J.A. Diez, C.U. Moulines, *Fundamentos de Filosofia de la Ciencia.* Barcelona: Ariel, 1997.

Henryk Ogryzko-Wiewiórowski is director of the Methods and Research Techniques Section at the Institute of Sociology, Department of Philosophy and Sociology at the Maria Curie-Skłodowska University,

Lublin, Poland. His research interests include the formal foundations of methodology of social sciences, research methods in sociology and the theoretical foundations of explanation in social sciences. His main publication is *Problem racjonalności wyboru społecznego* (The Problem of Rationality of Social Choice). Lublin: Wydawnictwo UMCS, 1986.

POZNAŃ STUDIES IN THE PHILOSOPHY OF THE SCIENCES AND THE HUMANITIES

Contents of back issues

VOLUME 1 (1975)

Main topics:
The Method of Humanistic Interpretation; The Method of Idealization; The Reconstruction of Some Marxist Theories.
(sold out)

VOLUME 2 (1976)

Main topics:
Idealizational Concept of Science; Categorial Interpretation of Dialectics.
(sold out)

VOLUME 3 (1977)

Main topic:
Aspects of the Production of Scientific Knowledge.
(sold out)

VOLUME 4 (1978)

Main topic:
Aspects of the Growth of Science.
(sold out)

VOLUME 5 (1979)

Main topic:
Methodological Problems of Historical Research.
(sold out)

VOLUME 6 (1982)

SOCIAL CLASSES ACTION & HISTORICAL MATERIALISM

Main topics:
On Classes; On Action; The Adaptive Interpretation of Historical Materialism; Contributions to Historical Materialism.
(sold out)

VOLUME 7 (1982)

DIALECTICAL LOGICS FOR THE POLITICAL SCIENCE
(Edited by Hayward R. Alker, Jr.)

VOLUME 8 (1985)

CONSCIOUSNESS: METHODOLOGICAL AND PSYCHOLOGICAL APPROACHES
(Edited by Jerzy Brzeziński)

VOLUME 9 (1986)

THEORIES OF IDEOLOGY AND IDEOLOGY OF THEORIES
(Edited by Piotr Buczkowski and Andrzej Klawiter)

VOLUME 10 (1987)

WHAT IS CLOSER-TO-THE-TRUTH?
A PARADE OF APPROACHES TO TRUTHLIKENESS
(Edited by Theo A.F. Kuipers)

VOLUME 11 (1988)

NORMATIVE STRUCTURES OF THE SOCIAL WORLD
(Edited by Giuliano di Bernardo)

VOLUME 12 (1987)

POLISH CONTRIBUTIONS TO THE THEORY AND PHILOSOPHY OF LAW
(Edited by Zygmunt Ziembiński)

VOLUME 13 (1988)

DIMENSIONS OF THE HISTORICAL PROCESS
(Edited by Leszek Nowak)

VOLUME 14 (1990)

ACTION AND PERFORMANCE: MODELS AND TESTS.
CONTRIBUTIONS TO THE QUANTITATIVE PSYCHOLOGY
AND ITS METHODOLOGY
(Edited by Jerzy Brzeziński and Tadeusz Marek)

VOLUME 15 (1989)

VISIONS OF CULTURE AND THE MODELS OF CULTURAL SCIENCES
(Edited by Jerzy Kmita and Krystyna Zamiara)

VOLUME 16 (1990)

IDEALIZATION I: GENERAL PROBLEMS
(Edited by Jerzy Brzeziński, Francesco Coniglione, Theo A.F. Kuipers and
Leszek Nowak)

VOLUME 17 (1990)

IDEALIZATION II: FORMS AND APPLICATIONS
(Edited by Jerzy Brzeziński, Francesco Coniglione, Theo A.F. Kuipers and
Leszek Nowak)

VOLUME 18 (1990)

STUDIES ON MARIO BUNGE'S *TREATISE*
(Edited by Paul Weingartner and Georg J.W. Dorn)

VOLUME 19 (1990)

NARRATION AND EXPLANATION.
CONTRIBUTION TO THE METHODOLOGY OF THE HISTORICAL RESEARCH
(Edited by Jerzy Topolski)

VOLUME 20 (1990)

Jürgen Ritsert
MODELS AND CONCEPTS OF IDEOLOGY

VOLUME 21 (1991)

PROBABILITY AND RATIONALITY
STUDIES ON L. JONATHAN COHEN'S PHILOSOPHY OF SCIENCE
(Edited by Ellery Eells and Tomasz Maruszewski)

VOLUME 22 (1991)

THE SOCIAL HORIZON OF KNOWLEDGE
(Edited by Piotr Buczkowski)

VOLUME 23 (1991)

ETHICAL DIMENSIONS OF LEGAL THEORY
(Edited by Wojciech Sadurski)

VOLUME 24 (1991)

ADVANCES IN SCIENTIFIC PHILOSOPHY
ESSAYS IN HONOUR OF PAUL WEINGARTNER ON THE OCCASION OF THE 60TH ANNIVERSARY OF HIS BIRTHDAY
(Edited by Gerhard Schurz and Georg J.W. Dorn)

VOLUME 25 (1992)

IDEALIZATION III: APPROXIMATION AND TRUTH
(Edited by Jerzy Brzeziński and Leszek Nowak)

VOLUME 26 (1992)

IDEALIZATION IV: INTELLIGIBILITY IN SCIENCE
(Edited by Craig Dilworth)

VOLUME 27 (1992)

Ryszard Stachowski

THE MATHEMATICAL SOUL.
AN ANTIQUE PROTOTYPE OF THE MODERN MATEMATISATION OF PSYCHOLOGY

VOLUME 28 (1993)

POLISH SCIENTIFIC PHILOSOPHY:
THE LVOV-WARSAW SCHOOL
(Edited by Francesco Coniglione, Roberto Poli and Jan Woleński)

VOLUME 29 (1993)

Zdzisław Augustynek and Jacek J. Jadacki

POSSIBLE ONTOLOGIES

VOLUME 30 (1993)

GOVERNMENT: SERVANT OR MASTER?
(Edited by Gerard Radnitzky and Hardy Bouillon)

VOLUME 31 (1993)

CREATIVITY AND CONSCIOUSNESS.
PHILOSOPHICAL AND PSYCHOLOGICAL DIMENSIONS
(Edited by Jerzy Brzeziński, Santo Di Nuovo, Tadeusz Marek and Tomasz Maruszewski)

VOLUME 32 (1993)

FROM ONE-PARTY-SYSTEM TO DEMOCRACY
(Edited by Janina Frentzel-Zagórska)

J. Frentzel-Zagórska, *Introduction*. **Part I: Theoretical Approaches** – Z. Bauman, *A Post-modern Revolution*; L. Holmes, *On Communism, Post-communism, Modernity and Post-modernity*; L. Nowak, *The Totalitarian Approach and the History of Socialism*; J. Pakulski, *East European Revolutions and 'Legitimacy Crisis'*. **Part II: The Transitional Period** – A. Czarnota, M. Krygier, *From State to Legal Traditions? Prospects for the Rule of Law after Communism*; M. Szabó, *Social Protest in a Post-communist Democracy: Taxi Drivers' Demonstration in Hungary*; Z. Bauman, *Dismantling Patronage State*; E. Mokrzycki, *Between Reform and Revolution: Eastern Europe Two Years after the Fall of Communism*; J. Frentzel-Zagórska, *The Road to Democratic Political System in Post-communist Eastern Europe*. **Part III: The Case of Yugoslavia** – R.F. Miller, *Yugoslavia: The End of the Experiment*.

VOLUME 33 (1993)

SOCIAL SYSTEM, RATIONALITY, AND REVOLUTION
(Edited by Marcin Paprzycki and Leszek Nowak)

Introduction. **On the Nature of Social System** – U. Preuss, *Political Order and Democracy. Carl Schmitt and his Influence*; K. Paprzycka, *A Paradox in Hobbes' Philosophy of Law*; S. Esquith, *Democratic Political Dialogue*; E. Jeliński, *Democracy in Polish Reformist Socialist Thought*; K. Paprzycka, *The Master and Slave Configuration in Hegel's System*; M. Godelier, *Lévi-Strauss, Marx and After. A Reappraisal of Structuralist and Marxist Tools for Social Logics*; K. Niedźwiadek, *On the Structure of Social System*; W. Czajkowski, *Social Being and Its Reproduction.* **On Rationality and Captivity** – M. Ziółkowski, *Power and Knowledge*, L. Nowak, *Two Inter-Human Limits to the Rationality of Man*; M. Paprzycki, *The Non-Christian Model of Man: An Attempt at Psychological Explanation*; R. Egiert, *Toward the Sophisticated Rationalistic Model of Man.* **On Social Revolution** – L. Nowak, *Revolution is an Opaque Progress but a Progress Nonetheless*; K. Paprzycka, M. Paprzycki, *How do Enslaved People Make Revolutions?*; G. Tomczak, *Is it Worth Winning a Revolution?*; K. Brzechczyn, *Civil Loops and the Absorption of Elites*; R. McCleary, *What Makes Marxist Historical Materialism Objective?*; G. Kotlarski, *Classes and Masses in Social Philosophy of Rosa Luxemburg.* **On Real Socialism** – E. Gellner, *The Civil and the Sacred*; W. Marciszewski, *Economics and the Idea of Information. Why Socialism must have Collapsed?*; L. Nowak, K. Paprzycka, M. Paprzycki, *On Multilinearity of Socialism*; A. Siegel, *The Overrepression Cycle in the Soviet Union. An Operationalization of a Theoretical Model*; K. Brzechczyn, *The State of the Teutonic Order as a Socialist Society.* **Discussions** – R. McCleary, *Socioanalysis and Philosophy*; W. Heller, *Methodological Remarks on the Public and the Private in Hannah Arendt's Political Philosophy*; K. Brzechczyn, *On Unsuccessful Conquest and Successful Subordination.*

VOLUME 34 (1994)

Izabella Nowakowa

IDEALIZATION V: THE DYNAMICS OF IDEALIZATIONS

Introduction; Chapter I: *Idealization and Theories of Correspondence*; Chapter II: *Dialectical Correspondence of Scientific Laws*; Chapter III: *Dialectical Correspondence in Science: Some Examples*; Chapter IV: *Dialectical Correspondence of Scientific Theories*; Chapter V: *Generalizations of the Rule of Correspondence*; Chapter VI: *Extensions of the Rule of Correspondence*; Chapter VII: *Correspondence and the Empirical Environment of a Theory*; Chapter VIII: *Some Methodological Problems of Dialectical Correspondence.*

VOLUME 35 (1993)

EMPIRICAL LOGIC AND PUBLIC DEBATE.
ESSAYS IN HONOUR OF ELSE M. BARTH
(Edited by E.C.W. Krabbe, R.J. Dalitz, and P.A. Smit)

Part I: Interpersonal Reasoning: Conflicts and Fallacies – T. Govier, *Needing Each Other for Knowledge: Reflections on Trust and Testimony*; J. Woods, *'Secundum quid' as a Research Programme*; G. Nuchelmans, *On the Fourfold Root of the 'argumentum ad hominem'*; F.H. van Eemeren, R. Grootendorst, *The History of the 'argumentum ad hominem' Since the Seventeenth Century*; L.S. van Epenhuysen, *Debate in a Bermuda Triangle of Medical Ethics*; E.C.W. Krabbe, *Reasonable Argument and Fallacies in the Kok-*

Stekelenburg Debate; **Part II: Linguistic and Conceptual Tools** – J.D. North, *Some Weak Links in the Great Chain of Being*; A. Næss, *'You assert this?' An Empirical Study of Weight Expressions*; R. Wiche, *Gerrit Mannoury on the Communicative Functions of Negation in Ordinary Language*; J. Hoepelman, T. van Hoof, *Default and Dogma*; Ch. Goossens, *On the Logic of Nonmoral Commitment*; W. Marciszewski, *Arguments Founded on Creative Definitions*; R. Jorna, *Cognitive Science and Connectivism: Friend and Enemy or Move and Counter-Move, an Application of Empirical Logic*; **Part III: Dialectical Climates and Tempests** – R.H. Johnson, *Dialectical Fields and Manifest Rationality*; P. du Preez, *Reason Which Cannot Be Reasoned With: What Is Public Debate and How Does it Change?*; M.A. Finocchiaro, *Logic, Democracy, and Mosca*; P.A. Smit, *The Logic of Virtue and Terror*; G. van Benthem van den Bergh, *On Obstacles to Public Debate*; J.P. van Bendegem, *Real-Life Mathematics versus Ideal Mathematics: The Ugly Truth*; **Part IV: The Disempowernment of Woman: Strategies and Counter-Moves** – V. Songe-Moller, *The Road of Being and the Exclusion of the Feminine. An Analysis of the Poem of Parmenides*; R.J. Dalitz, *The Subjection of Women in the Contractual Society. An Analysis of Thomas Hobbes' Theory of Agreement*; H. Schröder, *Anti-Semitism and anti-Feminism Again: The Dissemination of Otto Weininger's* Sex and Character *in the Seventies and Eighties*; J.R. Richards, *Traditional Spheres and Traditional Logic.*

VOLUME 36 (1994)

MARXISM AND COMMUNISM: POSTHUMOUS REFLECTIONS ON POLITICS, SOCIETY, AND LAW

(Edited by Martin Krygier)

M. Krygier, *Introduction*; A. Flis, *From Marx to Real Socialism: The History of a Utopia*; P. Marciniak, *The Collapse of Communism: Defeat or Opportunity for Marxism in Eastern Europe*; J. Clark, A. Wildavsky, *Chronicle of a Collapse Foretold: How Marx Predicted the Demise of Communism (Although He Called It "Capitalism")*; L. Nowak, *Political Theory and Socialism. On the Main Paradigms of Political Power and Their Methodological and Historical Legitimation*; E. Mokrzycki, *Marxism, Sociology, and "Real Socialism"*; R. Bäcker, *The Collapse of Communism and Theoretical Models*; A. Zybertowicz, *Three Deaths an Ideology: The Withering Away of Marxism and the Collapse of Communism. The Case of Poland*; M. Krygier, *Marxism, Communism, and the Rule of Law*; A. Czarnota, *Marxism, Ideology, and Law*; G. Skąpska, *The Legacy of Anti-Legalism*; A. Sajo, *Law and the Legal Scholarship in the Happiest Barrack and Among the Hungry Liberated: Personal Recollections.*

VOLUME 37 (1994)

THE SOCIAL PHILOSOPHY OF AGNES HELLER

(Edited by John Burnheim)

J. Burnheim, *Introduction*; M. Vajda, *A Lover of Philosophy – A Lover of Europe*; P. Despoix, *On the Possibility of a Philosophy of Values. A Dialogue within the Budapest School*; M. Jay, *Women in Dark Times: Agnes Heller and Hannah Arendt*; J.P. Arnason, *The Human Condition and the Modern Predicament*; R.J. Bernstein, *Agnes Heller: Philosophy, Rational Utopia and Praxis*; Z. Bauman, *Narrating Modernity*; P. Beilharz, *Theories of History – Agnes Heller and R.G. Collingwood*; P. Wolin, *Heller's Theory of Everyday Life*; P. Harrison, *Radical Philosophy and the Theory of Modernity*; A.J. Jacobson, *The Limits of Formal Justice*; P. Murphy, *Civility and Radicalism*; P. Murphy, *Pluralism and Politics*; V. Camps, *The Good Life: A Moral Gesture*; L. Boella, *Philosophy Beyond the Baseless and Tragic Character of Action*; G. Márkus, *The Politics of Morals*; A. Heller, *A Reply to My Critics*; *The Bibliography of Agnes Heller.*

VOLUME 38 (1994)

IDEALIZATION VI: IDEALIZATION IN ECONOMICS
(Edited by Bert Hamminga and Neil B. De Marchi)

Introduction – B. Hamminga, N. De Marchi, *Preface*; B. Hamminga, N. De Marchi, *Idealization and the Defence of Economics: Notes Toward a History*. **Part I: General Observations on Idealization in Economics** – K.D. Hoover, *Six Queries about Idealization in an Empirical Context*; B. Walliser, *Three Generalization Processes for Economic Models*; S. Cook, D. Hendry, *The Theory of Reduction in Econometrics*; M.C.W. Janssen, *Economic Models and Their Applications*; A.G. de la Sienra, *Idealization and Empirical Adequacy in Economic Theory*; I. Nowakowa, L. Nowak, *On Correspondence between Economic Theories*; U. Mäki, *Isolation, Idealization and Truth in Economics*. **Part II: Case Studies of Idealization in Economics** – N. Cartwright, *Mill and Menger: Ideal Elements and Stable Tendencies*; W. Balzer, *Exchange Versus Influence: A Case of Idealization*; K. Cools, B. Hamminga, T.A.F. Kuipers, *Truth Approximation by Concretization in Capital Structure Theory*; D.M. Hausman, *Paul Samuelson as Dr. Frankenstein: When an Idealization Runs Amuck*; H.A. Keuzenkamp, *What if an Idealization is Problematic? The Case of the Homogeneity Condition in Consumer Demand*; W. Diederich, *Nowak on Explanation and Idealization in Marx's Capital*; G. Jorland, *Idealization and Transformation*; J. Birner, *Idealizations and Theory Development in Economics. Some History and Logic of the Logic Discovery*. **Discussions** – L. Nowak, *The Idealizational Methodology and Economics. Replies to Diederich, Hoover, Janssen, Jorland and Mäki*.

VOLUME 39 (1994)

PROBABILITY IN THEORY-BUILDING.
EXPERIMENTAL AND NON-EXPERIMENTAL APPROACHES
TO SCIENTIFIC RESEARCH IN PSYCHOLOGY
(Edited by Jerzy Brzeziński)

Part I: Probability and the Idealizational Theory of Science – M. Gaul, *Statistical Dependencies, Statements and the Idealizational Theory of Science*. **Part II: Probability – Theoretical Concepts in Psychology – Measurement** – D. Wahlstein, *Probability and the Understanding of Individual Differences*; B. Krause, *Modeling Cognitive Learning Steps*; D. Heyer, R. Mausfeld, *A Theoretical and Experimental Inquiry into the Relation of Theoretical Concepts and Probabilistic Measurement Scales in Experimental Psychology*. **Part III: Methods of Data Analysis** – T.B. Iwiński, *Rough Set Methods in Psychology*; W. Koutstaal, R. Rosenthal, *Contrast Analysis in Behavioral Research*. **Part IV: Artifacts in Psychological Research and Diagnostic Assessment** – D.B. Strohmetz, R.L. Rosnow, *A Mediational Model of Research Artifacts*; J. Brzeziński, *Dimensions of Diagnostic Space*.

VOLUME 40 (1995)

THE HERITAGE OF KAZIMIERZ AJDUKIEWICZ
(Edited by Vito Sinisi and Jan Woleński)

Preface; J. Giedymin, *Ajdukiewicz's Life and Personality*; K. Ajdukiewicz, *My Philosophical Ideas*; L. Albertazzi, *Some Elements of Transcendentalism in Ajdukiewicz*; T. Batóg, *Ajdukiewicz and the Development of Formal Logic*; A. Church, *A Theory of the Meaning of Names*; M. Czarniawska, *The Way from Concept to Thought. Does it Exist in Ajdukiewicz's Semantical Theory?*; E. Dölling, *Real Objects and Existence*; J. Giedymin, *Radical Conventionalism, Its Background and Evolution: Poincaré, Le Roy, Ajdukiewicz*; J.J.

Jadacki, *Definition, Explication and Paraphrase in the Ajdukiewiczian Tradition*; G. Küng, *Ajdukiewicz's Contribution to the Realism/Idealism Debate*; J. Maciaszek, *Problems of Proper Names: Ajdukiewicz and Some Contemporary Results*; W. Marciszewski, *Real Definitions and Creativity*; K. Misiuna, *Categorial Grammar and Ontological Commitment*; L. Nowak, *Ajdukiewicz and the Status of the Logical Theory of Natural Language*; A. Olech, *Several Remarks on Ajdukiewicz's and Husserl's Approaches to Meaning*; A. Orenstein, *Existence Sentences*; J. Pasek, *Ajdukiewicz on Indicative Conditionals*; R. Poli, *The Problem of Position: Ajdukiewicz and Leibniz on Intensional Expressions*; M. Przełęcki, *The Law of Excluded Middle and the Problem of Idealism*; H. Skolimowski, *Ajdukiewicz, Rationality and Language*; K. Szaniawski, *Ajdukiewicz on Non-deductive Inference*; A. Varzi, *Variable-Binders as Functors*; V. Vasyukov, *Categorial Semantics for Ajdukiewicz-Lambek Calculus*; H. Wassing, *On the Expressiveness of Categorial Grammar*; J. Woleński, *On Ajdukiewicz's Refutation of Scepticism*; *Bibliography of Ajdukiewicz*; *Bibliography on Ajdukiewicz (writings in western languages)*.

VOLUME 41 (1994)

HISTORIOGRAPHY BETWEEN MODERNISM AND POSTMODERNISM. CONTRIBUTIONS TO THE METHODOLOGY OF THE HISTORICAL RESEARCH
(Edited by Jerzy Topolski)

Editor's Introduction; J. Topolski, *A Non postmodernist Analysis of Historical Narratives*; F.R. Ankersmit, *Historism, Postmodernism and Historiography*; D. Carr, *Getting the Story Straight: Narrative and Historical Knowledge*; W. Wrzosek, *The Problem of Cultural Imputation in History. Culture versus History*; J. Tacq, *Causality and Virtual Finality*; G. Zalejko, *Soviet Historiography as a "Normal Science"*; H. Mamzer and J. Ostoja-Zagórski, *Deconstruction of Evolutionist Paradigm in Archaeology*; N. Lautier, *At the Crossroad of Epistemology and Psychology: Prospects of a Didactic of History*; Teresa Kostyrko, *Remarks on "Aesthetization" in Science on the Basis of History*.

VOLUME 42 (1995)

IDEALIZATION VII: IDEALIZATION, STRUCTURALISM, AND APPROXIMATION
(Edited by Martti Kuokkanen)

Idealization, Approximation and Counterfactuals in the Structuralist Framework – T.A.F. Kuipers, *The Refined Structure of Theories*; C.U. Moulines and R. Straub, *Approximation and Idealization from the Structuralist Point of View*; I.A. Kieseppä, *A Note on the Structuralist Account of Approximation*; C.U. Moulines and R. Straub, *A Reply to Kieseppä*; W. Balzer and G. Zoubek, *Structuralist Aspects of Idealization*; A. Ibarra and T. Mormann, *Counterfactual Deformation and Idealization in a Structuralist Framework*; I.A. Kieseppä, *Assessing the Structuralist Theory of Verisimilitude*. **Idealization, Approximation and Theory Formation** – L. Nowak, *Remarks on the Nature of Galileo's Methodological Revolution*; I. Niiniluoto, *Approximation in Applied Science*; E. Heise, P. Gerjets and R. Westermann, *Idealized Action Phases. A Concise Rubicon Theory*; K.G. Troitzsch, *Modelling, Simulation, and Structuralism*; V. Rantala and T. Vadén, *Idealization in Cognitive Science. A Study in Counterfactual Correspondence*; M. Sintonen and M. Kiikeri, *Idealization in Evolutionary Biology*; T. Tuomivaara, *On Idealizations in Ecology*; M. Kuokkanen and M. Häyry, *Early Utilitarianism and Its Idealizations from a Systematic Point of View*. **Idealization, Approximation and Measurement** – R. Westermann, *Measurement-Theoretical Idealizations and Empirical Research Practice*; U. Konerding,

Probability as an Idealization of Relative Frequency. A Case Study by Means of the BTL-Model; R. Suck and J. Wienöbst, *The Empirical Claim of Probality Statements, Idealized Bernoulli Experiments and their Approximate Version*; P.J. Lahti, *Idealizations in Quantum Theory of Measurement.*

VOLUME 43 (1995)

Witold Marciszewski and Roman Murawski
MECHANIZATION OF REASONING IN A HISTORICAL PERSPECTIVE

Chapter 1: *From the Mechanization of Reasoning to a Study of Human Intelligence*; Chapter 2: *The Formalization of Arguments in the Middle Ages*; Chapter 3: *Leibniz's Idea of Mechanical Reasoning at the Historical Background*; Chapter 4: *Between Leibniz and Boole: Towards the Algebraization of Logic*; Chapter 5: *The English Algebra of Logic in the 19th Century*; Chapter 6: *The 20th Century Way to Formalization and Mechanization*; Chapter 7: *Mechanized Deduction Systems.*

VOLUME 44 (1995)

THEORIES AND MODELS IN SCIENTIFIC PROCESSES
(Edited by William Herfel, Władysław Krajewski,
Ilkka Niiniluoto and Ryszard Wójcicki)

Introduction; **Part 1. Models in Scientific Processes** – J. Agassi, *Why there is no Theory of Models?*; M. Czarnocka, *Models and Symbolic Nature of Knowledge*; A. Grobler, *The Representational and the Non-Representational in Models of Scientific Theories*; S. Hartmann, *Models as a Tool for the Theory Construction: Some Strategies of Preliminary Physics*; W. Herfel, *Nonlinear Dynamical Models as Concrete Construction*; E. Kałuszyńska, *Styles of Thinking*; S. Psillos, *The Cognitive Interplay Between Theories and Models: the Case of 19th Century Optics.* **Part 2. Tools of Science** – N.D. Cartwright, T. Shomar, M. Suarez, *The Tool-Box of Science*; J. Echeverria, *The Four Contexts of Scienctific Activity*; K. Havas, *Continuity and Change: Kinds of Negation in Scientific Progress*; M. Kaiser, *The Independence of Scientific Phenomena*; W. Krajewski, *Scientific Meta-Philosophy*; I. Niiniluoto, *The Emergence of Scientific Specialties: Six Models*; L. Nowak, *Antirealism, (Supra-) Realism and Idealization*; R.M. Nugayev, *Classic, Modern and Postmodern Scientific Unification*; V. Rantala, *Translation and Scientific Change*; G. Schurz, *Theories and Their Applications – a Case of Nonmonotonic Reasoning*; W. Strawiński, *The Unity of Science Today*; V. Torosian, *Are the Ethics and Logic of Science Compatible?* **Part 3. Unsharp Approaches in Science** – E.W. Adams, *Problems and Prospects in a Theory of Inexact First-Order Theories*; W. Balzer, G. Zoubek, *On the Comparision of Approximative Empirical Claims*; G. Cattaneo, M. Luisa Dalla Chiara, R. Giuntini, *The Unsharp Approaches to Quantum Theory*; T.A.F. Kuipers, *Falsification Versus Effcient Truth Approximation*; B. Lauth, *Limiting Decidability and Probability*; J. Pykacz, *Many-Valued Logics in Foundations of Quantum Mechanics*; R.R. Zapatrin, *Logico-Algebraic Approach to Spacetime Quantization.*

VOLUME 45 (1995)

COGNITIVE PATTERNS IN SCIENCE AND COMMON SENSE. GRONINGEN STUDIES IN PHILOSOPHY OF SCIENCE, LOGIC, AND EPISTEMOLOGY
(Edited by Theo A.F. Kuipers and Anne Ruth Mackor)

L. Nowak, *Foreword*; **General Introduction** – T.A.F. Kuipers and A.R. Mackor, *Cognitive Studies of Science and Common Sense*; **Part I: Conceptual Analysis in Service of Various Research Programmes** – H. Zandvoort, *Concepts of Interdisciplinarity and Environmental Science*; R. Vos, *The Logic and Epistemology of the Concept of Drug and Disease Profile*; R.C. Looijen, *On the Distinction Between Habitat and Niché, and Some Implications for Species' Differentiation*; G.J. Stavenga, *Cognition, Irreversibility and the Direction of Time*; R. Dalitz, *Knowledge, Gender and Social Bias*; **Part II: The Logic of the Evaluation of Arguments, Hypotheses, Rules, and Interesting Theorems** – E.G.W. Krabbe, *Can We Ever Pin One Down to a Formal Fallacy?*; T.A.F. Kuipers, *Explicating the Falsificationist and Instrumentalist Methodology by Decomposing the Hypothetico-Deductive Method*; A. Keupink, *Causal Modelling and Misspecification: Theory and Econometric Historical Practice*; M.C.W. Janssen and Y.-H. Tan, *Default Reasoning and Some Applications in Economics*; B. Hamminga, *Interesting Theorems in Economics*; **Part III: Three Challenges to the Truth Approximation Programme** – S.D. Zwart, *A Hidden Variable in the Discussion About 'Language Dependency' of Truthlikeness*; H. Hettema and T.A.F. Kuipers, *Sommerfeld's* Atombau: *A Case Study in Potential Truth Approximation*; R. Festa, *Verisimilitude, Disorder, and Optimum Prior Probabilities*; **Part IV: Explicating Psychological Intuitions** – A.R. Mackor, *Intentional Psychology is a Biological Discipline*; J. Peijnenburg, *Hempel's Rationality. On the Empty Nature of Being a Rational Agent*; L. Guichard, *The Causal Efficacy of Propositional Attitudes*; M. ter Hark, *Connectionism, Behaviourism and the Language of Thought.*

VOLUME 46 (1996)

POLITICAL DIALOGUE: THEORIES AND PRACTICE
(Edited by Stephen L. Esquith)

Introduction – S.L. Esquith, *Political Dialogue and Political Virtue.* **Part I: The Modern Clasics** – A.J. Damico, *Reason's Reach: Liberal Tolerance and Political Discourse*; T.R. Machan, *Individualism and Political Dialogue*; R. Kukla, *The Coupling of Human Souls: Rousseau and the Problem of Gender Relations*; D.F. Koch, *Dialogue: An Essay in the Instrumentalist Tradition.* **Part II: Toward a Democratic Synthesis** – E. Simpson, *Forms of Political Thinking and the Persistence of Practical Philosophy*; J.B. Sauer, *Discoursee, Consensus, and Value: Conversations about the Intelligible Relation Between the Private and Public Spheres*; M. Kingwell, *Phronesis and Political Dialogue*, R.T. Peterson, *Democracy and Intellectual Mediation – After Liberalism and Socialism.* **Part III: Dialogue in Practice** – S. Rohr Scaff, L.A. Scaff, *Political Dialogue in the New Germany: The Burdens of Culture and an Asymmetrical Past*; J.H. Read, *Participation, Power, and Democracy*; S.E. Bennett, B. Fisher, D. Resnick, *Speaking of Politics in the United States: Who Talks to Whom, Why, and Why Not*; J. Forester, *Beyond Dialogue to Transformative Learning: How Deliberative Rituals Encourage Political Judgement in Community Planning Processes*; A. Fatić, *Retribution in Democracy.*

VOLUME 47 (1996)

EPISTEMOLOGY AND HISTORY. HUMANITIES AS A PHILOSOPHICAL PROBLEM AND JERZY KMITA'S APPROACH TO IT
(Edited by Anna Zeidler-Janiszewska)

VOLUME 48 (1996)

THE SOCIAL PHILOSOPHY OF ERNEST GELLNER
(Edited by John A. Hall and Ian Jarvie)

Modernity: Reflection on the Theory of Nationalism; M. Mann, *The Emergence of Modern European Nationalism*; N. Stagardt, *Gellner's Nationalism: The Spirit of Modernisation?* **Part 3: Patterns of Development** – P. Burke, *Reflections on the History of Encyclopaedias*; A. MacFarlane, *Ernest Gellner and the Escape to Modernity*; R. Dore, *Soverein Individuals*; S. Eisenstadt, *Japan: Non Axial Modernity*; M. Ferro, *L'Indépendance Telescopée: De la Décolonisation a L'Impérialisme Multinational.* **Part 4: Islam** – A. Hammoudi, *Segmentarity, Social stratification, Political Power and Sainthood: Reflections on Gellner's Theses*; H. Munson, Jr., *Rethinking Gellner's Segmentary Analysis of Morocco's Ait'Atta*; J. Baechler, *Sur le charisme*; Ch. Lindholm, *Despotism and Democracy: State and Society in PreModern Middle East*; H. Munson, *Jr. Muslism and Jew in Morocco: Reflections on the Distinction between Belief and Behaviour*; T. Asad, *The Idea of an Anthropology of Islam.* **Part 5: Science and Disenchantment** – P. Anderson, *Science, Politics, Enchantment*; R. Schroeder, *From the Big Divide to the Rubber Cage: Gellner's Conception of Science and Technology*; J. Davis, *Irrationality in Social Life.* **Part 6: Relativism and Universals** – J. Skorupski, *The Post-Modern Hume: Ernest Gellner's 'Enlightenment Fundamentalism'*; J. Wettersten, *Ernest Gellner: A Wittgensteinian Rationalist*; I. Jarvie, *Gellner's Positivism*; R. Boudon, *Relativising Relativism: When Sociology Refutes the Sociology of Science*; R. Aya, *The Empiricist Exorcist.* **Part 7: Philosophy of History** – W. McNeill, *A Swang Song for British Liberalism?*; A. Park, *Gellner and the Long Trends of History*; E. Leone, *Marx, Gellner, Power*; R. Langlois, *Coercion, Cognition and Production: Gellner's Challenge to Historical Materialism and Post-Modernism*; E. Gellner, *Reply to my Critics*; I. Jarvie, *Complete Bibliography of Gellner's Work.*

VOLUME 49 (1996)

THE SIGNIFICANCE OF POPPER'S THOUGHT
(Edited by Stefan Amsterdamski)

Karl Popper's Three Worlds – J. Watkins, *World 1, World 2 and the Theory of Evolution*; A. Grobler, *World 3 and the Cunning of Reason.* **The Scientific Method as Ethics** – J. Agassi, *Towards Honest Public Relations of Science*; S. Amsterdamski, *Between Relativism and Absolutism: the Popperian Ideal of Knowledge.* **The Open Society and its Prospects** – E. Gellner, *Karl Popper – The Thinker and the Man*; J. Woleński, *Popper on Prophecies and Predictions.*

VOLUME 50 (1996)

THE IDEA OF THE UNIVERSITY
(Edited by Jerzy Brzeziński and Leszek Nowak)

Introduction – K. Twardowski, *The Majesty of the University.* **I** – Z. Ziembiński, *What Can be Saved of the Idea of the University?*; L. Kołakowski, *What Are Universities for?*; L. Gumański, *The Ideal University and Reality*; Z. Bauman, *The Present Crisis of the Universities.* **II** – K. Ajdukiewicz, *On Freedom of Science*; H. Samsonowicz, *Universities and Democracy*; J. Topolski, *The Commonwealth of Scholars and New Conceptions of Truth*; K. Szaniawski, *Plus ratio quam vis.* **III** – L. Koj, *Science, Teaching and Values*; K. Szaniawski, *The Ethics of Scientific Criticism*; J. Brzeziński, *Ethical Problems of Research Work of Psychologists.* **IV** – J. Goćkowski, *Tradition in Science*; J. Kmita, *Is a "Creative Man of Knowledge" Needed in University Teaching?*; L. Nowak, *The Personality of Researchers and the Necessity of Schools in Science.* **Recapitulation** – J. Brzeziński, *Reflections on the University.*

VOLUME 51 (1997)

KNOWLEDGE AND INQUIRY: ESSAYS ON JAAKKO HINTIKKA'S EPISTEMOLOGY AND PHILOSOPHY OF SCIENCE
(Edited by Matti Sintonen)

M. Sintonen, *From the Science of Logic to the Logic of Science*. **I: Historical Perspectives** – Z. Bechler, *Hintikka on Plentitude in Aristotle*; M.-L. Kakkuri-Knuuttila, *What Can the Sciences of Man Learn from Aristotle?*; M. Kusch, *Theories of Questions in German-Speaking Philosophy Around the Turn of the Century*; N.-E. Sahlin, *'He is no Good for My Work': On the Philosophical Relations between Ramsey and Wittgenstein*. **II: Formal Tools: Induction, Observation and Identifiability** – T.A.F. Kuipers, *The Carnap-Hintikka Programme in Inductive Logic*; I. Levi, *Caution and Nonmonotic Inference*; I. Niiniluoto, *Inductive Logic, Atomism, and Observational Error*; A. Mutanen, *Theory of Identifiability*. **III: Questions in Inquiry: The Interrogative Model** – S. Bromberger, *Natural Kinds and Questions*; S.A. Kleiner, *The Structure of Inquiry in Developmental Biology*; A. Wiśniewski, *Some Foundational Concepts of Erotetic Semantics*; J. Woleński, *Science and Games*. **IV: Growth of Knowledge: Explanation and Discovery** – M. Sintonen, *Explanation: The Fifth Decade*; E. Weber, *Scientific Explanation and the Interrogative Model of Inquiry*; G. Gebhard, *Scientific Discovery, Induction, and the Multi-Level Character of Scientific Inquiry*; M. Kiikeri, *On the Logical Structure of Learning Models*. **V: Jaakko Hintikka: Replies**.

VOLUME 52 (1997)

Helena Eilstein
LIFE CONTEMPLATIVE, LIFE PRACTICAL.
AN ESSAY ON FATALISM

Preface. **Chapter One: Oldcomb and Newcomb** – 1. *In the King Comb's Chamber of Game*; 2. *The Newcombian Predicaments*. **Chapter Two: Ananke** – 1. *Fatalism: What It Is Not?*; 2. *Fatalism: What Is It?*; 3. *Fatalism and* a priori *Arguments*; 4. *Fatalism and 'Internal' Experience*; 5. *Determinism, Indeterminism and Fatalism*; 6. *Transientism, Eternism and Fatalism*; 7. *Fatalism: What It Does Not Imply?*. **Chapter Three: Fated Freedom** – 1. *More on Libertarianism*; 2. *On the Deterministic Concept of Freedom*; 3. *Moral Self and Responsibility in the Light of Probabilism*; 4. *Fated Freedom*. **Chapter Four: The Virus of Fatalism** – 1. *Fatalism and Problems of Cognition*; 2. *The Virus of Fatalism: Why Mostly Harmless?*

VOLUME 53 (1997)

Dimitri Ginev
A PASSAGE TO THE HERMENEUTIC PHILOSOPHY OF SCIENCE

Preface; **Introduction**: *Topics in the Hermeneutic Philosophy of Science*; **Chapter 0**: *On the Limits of the Rational Reconstruction of Scientific Knowledge*; **Chapter 1**: *On the Hermeneutic Nature of Group Rationality*; **Chapter 2**: *Towards a Hermeneutic Theory of Progressive Change in Scientific Development*; **Chapter 3**: *Beyond Naturalism and Traditionalism*; **Chapter 4**: *A Critical Note on Normative Naturalism*; **Chapter 5**: *Micro and Macrohermeneutics of Science*; *Concluding Remarks*.

VOLUME 54 (1997)

IN ITINERE. EUROPEAN CITIES AND THE BIRTH OF MODERN SCIENTIFIC PHILOSOPHY
(Edited by Roberto Poli)

Introduction – R. Poli, *In itinere. Pictures from Central-European Philosophy*. **Stages of the Tour** – K. Schuhmann, *Philosophy and Art in Munich around the Turn of the Century*; M. Libardi, *In itinere: Vienna 1870-1918*; W. Baumgartner, *Nineteenth-Century Würzburg: The Development of the Scientific Approach to Philosophy*; L. Dappiano, *Cambridge and the Austrian Connection*; J. Sebestik, *Prague Mosaic. Encounters with the Prague Philosophers*; J.J. Jadacki, *Warsaw: The Rise and Decline of Modern Scientific Philosophy in the Capital City of Poland*; J. Woleński, *Lvov*; L. Albertazzi, *Science and the Avant-Garde in Early Nineteenth-Century Florence*; F. Minazzi, *The Presence of Phenomenology in Milan between the Two World Wars. The Contribution of Antonio Banfi and Giulio Preti*.

VOLUME 55 (1997)

REALISM AND QUANTUM PHYSICS
(Edited by Evandro Agazzi)

Introduction – E. Agazzi. **Part One: Philosophical Considerations** – P. Horwich, *Realism and Truth*, E. Agazzi, *On the Criteria for Establishing the Ontological Status of Different Entities*; A. Baltas, *Constraints and Resistance: Stating a Case for Negative Realism*; M. Paty, *Predicate of Existence and Predicability for a Theoretical Object in Physics*. **Part Two: Observability and Hidden Entities** – F. Bonsak, *Atoms: Lessons of a History*; A. Cordero, *Arguing for Hidden Realities*; B. d'Espagnat, *On the Difficulties that Attributing Existence to "Hidden" Quantities May Raise*; M. Pauri, *The Quantum, Space-Time and Observation*. **Part Three: Applications to Quantum Physics** – D. Albert, *On the Phenomenology of Quantum-Mechanical Superpositions*; G.C. Ghiraldi, *Realism and Quantum Mechanics*; M. Crozon, *Experimental Evidence of Quark Structure Inside Hadrons*.

VOLUME 56 (1997)

IDEALIZATION VIII: MODELLING IN PSYCHOLOGY
(Edited by Jerzy Brzeziński, Bodo Krause and Tomasz Maruszewski)

Part I: Philosophical and Methodological Problems of Cognition Process – J. Wane, *Idealizing the Cartesian-Newtonian Paradigm as Reality: The Impact of New-Paradigm Physics on Psychological Theory*; E. Hornowska, *Operationalization of Psychological Magnitudes. Assumptions-Structure-Consequences*; T. Bachmann, *Creating Analogies – On Aspects of the Mapping Process between Knowledge Domains*; H. Schaub, *Modelling Action Regulation*. **Part II: The Structure of Ideal Learning Process** – S. Ohlson, J.J. Jewett, *Ideal Adaptive Agents and the Learning Curve*; B. Krause, *Towards a Theory of Cognitive Learning*; B. Krause, U. Gauger, *Learning and Use of Invariances: Experiments and Network Simulation*; M. Friedrich, *"Reaction Time" in the Neural Network Module ART 1*. **Part III: Control Processes in Memory** – J. Tzelgov, V. Yehene, M. Naveh-Benjamin, *From Memory to Automaticity and Vice Versa: On the Relation between Memory and Automaticity*; H. Hagendorf, S. Fisher, B. Sá, *The Function of Working Memory in Coordination of Mental Transformations*; L. Nowak, *On Common-Sense and (Para-) Idealization*; I. Nowakowa, *On the Problem of Induction. Toward an Idealizational Paraphrase*.

VOLUME 57 (1997)

EUPHONY AND LOGOS
ESSAYS IN HONOUR OF MARIA STEFFEN-BATÓG AND TADEUSZ BATÓG
(Edited by Roman Murawski, Jerzy Pogonowski)

Preface. **Scientific Works of Maria Steffen-Batóg and Tadeusz Batóg** – *List of Publications of Maria Steffen-Batóg*; *List of Publications of Tadeusz Batóg*; J. Pogonowski, *On the Scientific Works of Maria Steffen-Batóg*; Jerzy Pogonowski, *On the Scientific Works of Tadeusz Batóg*; W. Lapis, *How Should Sounds Be Phonemicized?*; P. Nowakowski, *On Applications of Algorithms for Phonetic Transcription in Linguistic Research*; J. Pogonowski, *Tadeusz Batóg's Phonological Systems.* **Mathematical Logic** – W. Buszkowski, *Incomplete Information Systems and Kleene 3-valued Logic*; M. Kandulski, *Categorial Grammars with Structural Rules*; M. Kołowska-Gawiejnowicz, *Labelled Deductive Systems for the Lambek Calculus*; R. Murawski, *Satisfaction Classes – a Survey*; K. Świrydowicz, *A New Approach to Dyadic Deontic Logic and the Normative Consequence Relation*; W. Zielonka, *More about the Axiomatics of the Lambek Calculus.* **Theoretical Linguistics** – J.J. Jadacki, *Troubles with Categorial Interpretation of Natural Language*; M. Karpiński, *Conversational Devices in Human-Computer Communication Using WIMP UI*; W. Maciejewski, *Qualitative Orientation and Gramatical Categories*; Z. Vetulani, *A System of Computer Understanding of Text*; A. Wójcik, *The Formal Development of van Sandt's Presupposition Theory*; W. Zabrocki, *Psychologism in Noam Chomsky's Theory (Tentative Critical Remarks)*; R. Zuber, *Defining Presupposition without Negation.* **Philosophy of Language and Methodology of Sciences** – J. Kmita, *Philosophical Antifundamentalism*; A. Luchowska, *Peirce and Quine: Two Views on Meaning*; S. Wiertlewski, *Method According to Feyerabend*; J. Woleński, *Wittgenstein and Ordinary Language*; K. Zamiara, *Context of Discovery – Context of Justification and the Problem of Psychologism.*

VOLUME 58 (1997)

THE POSTMODERNIST CRITIQUE OF THE PROJECT OF ENLIGHTENMENT
(Edited by Sven-Eric Liedman)

S.-E. Liedman, *Introduction*; R. Wokler, *The Enlightenment Project and its Critics*; M. Benedikt, *Die Gegenwartsbedeutung von Kants aufklärender Akzeptanz und Zurückweisung des Modells der Naturwissenschaft für zwischenmenschliche Verhältnisse: Verfehlte Beziehungen der Geisterwelt Swedenborgs*; S.-E. Liedman, *The Crucial Role of Ethics in Different Types of Enlightenment (Condorcet and Kant)*; S. Dahlstedt, *Forms of the Ineffable: From Kant to Lyotard*; P. Magnus Johansson, *On the Enlightenment in Psycho-Analysis*; E. Lundgren-Gothlin, *Ethics, Feminism and Postmodernism: Seyla Benhabib and Simone de Beauvoir*; E. Kiss, *Gibt es ein Projekt der Aufklärung und wenn ja, wie viele (Aufklärung vor dem Horizont der Postmoderne)*; E. Kennedy, *Enlightenment Anticipations of Postmodernist Epistemology*; L. Nowak, *On Postmodernist Philosophy: An Attempt to Identify its Historical Sense*; M. Castillo, *The Dilemmas of Postmodern Individualism.*

VOLUME 59 (1997)

BEYOND ORIENTALISM. THE WORK OF WILHELM HALBFASS AND
ITS IMPACT ON INDIAN AND CROSS-CULTURAL STUDIES
(Edited by Eli Franco and Karin Preisendanz)

E. Franco, K. Preisendanz, *Introduction and Editorial Essay on Wilhelm Halbfass*; Publications by Wilhelm Halbfass; W. Halbfass, *Research and Reflections: Responses to my*

Respondents. Beyond Orientalism? Reflections on a Current Theme. **Part I: Cross-Cultural Encounter and Dialogue** – F. X. Clooney, SJ, *Wilhelm Halbfass and the Openness of the Comparative Project*; F. Dallmayr, *Exit from Orientalism: Comments on Wilhelm Halbfass*; S. D. Serebriany, *Some Marginal Notes on India and Europe*; R. Sen (née Mookerjee), *Some Reflections on India and Europe: An Essay in Understanding*; K. Karttunen, *Greeks and Indian Wisdom*; D. Killingley, *Mlecchas, Yavanas and Heathens: Interacting Xenologies in Early Nineteenth-Century Calcutta*; W. Halbfass, *Research and Reflections: Responses to my Respondents. Cross-Cultural Encounter and Dialogue.* **Part II: Issues of Comparative Philosophy** – J. Nath Mohanty, *Between Indology and Indian Philosophy*; J. S. O'Leary, *Heidegger and Indian Philosophy*; S. Rao, *"Subordinate" or "Supreme"? The Nature of Reason in India and the West*; R. Iveković, *The Politics of Comparative Philosophy*; B.-A. Scharfstein, *The Three Philosophical Traditions*; W. Halbfass, *Research and Reflections: Responses to my Respondents. Issues of Comparative Philosophy.* **Part III: Topics in Classical Indian Philosophy** – J. E. M. Houben, *Bhartṛhari's Perspectivism: The Vṛtti and Bhartṛhari's Perspectivism in the First kāṇḍa of the Vākyapadīya*; J. Bronkhorst, *Philosophy and the Vedic Exegesis in the Mīmāṃsā*; J. Taber, *The Significance of Kumārila's Philosophy*; K. Harikai, *Kumārila's Acceptance and Modification of Categories of the Vaiśeṣika School*; V. Lysenko, *The Vaiśeṣika Notions of ākāsa and diś from the Perspective of Indian Ideas of Space*; B. M. Perry, *Early Nyāya and Hindu Orthodoxy: ānvīkṣikī and adhikāra*; W. Halbfass, *Research and Reflections: Responses to my Respondents: Topics in Classical Indian Philosophy.* **Part IV: Developments and Attitudes in Neo-Hinduism** – A. O. Fort, *Jīvanmukti and Social Service in Advaita and Neo-Vedānta*; S. Elkman, *Religious Plurality and Swami Vivekananda.* **Part V: Indian Religion, Past and Present** – M. Hara, *A Note on dharmasya sūkṣmā gatiḥ*; A. Wezler, *The Story of Aṇī-Māṇḍavya as told in the Mahābhārata: Its significance for Indian Legal and Religious History*; Y. Grinshpon, *Experience and Observation in Traditional and Modern Pātañjala Yoga*; F. J. Korom, *Language Belief and Experience in Bengali Folk Religion*; W. Halbfass, *Research and Reflection: Responses to my Respondents. Developments and Attitudes in Neo-Hinduism; Indian religion, Past and Present.*

VOLUME 60 (1998)

MARX'S THEORIES TODAY
(Edited by Ryszard Panasiuk and Leszek Nowak)

R. Panasiuk, *Introduction*; **Part I: On Dialectics and Ontology** – S.-E. Liedman, *Engels and the Laws of Dialectics*; R. Panasiuk, *On Dialectics in Marxism Again*; R. Albritton, *The Unique Ontology of Capital*; R. Washner, *It is not Singularity that Governs the Nature of Things. The Principle of Isolated Individual and Its Negation in Marx's Doctoral Thesis*; **Part II: On Historical Materialism and Social Theories** – Z. Cackowski, *The Continuing Validity of the Marxian Thought*; P. Casal, *From Unilineal to Universal Historical Materialism*; I. Hunt, *A Dialectical Interpretation and Resurrection of Historical Materialism*; W. Krajewski, *The Triumph of Historical Materialism*; L. Nowak, *The Adaptive Interpretation of Historical Materialism: A Survey. On a Contribution to Polish Analytical Marxism*; M. Kozłowski, *A New Look at Capitalism. Between the* Decommunisation *of Marx's and the* Defeudalisation *of Hegel's Visions of Capitalism*; F. Moseley, *An Empirical Appraisal of Marx's Economic Theory*; Ch. Bertram and A. Carling, *Stumbling into Revolution. Analytical Marxism, Rationality and Collective Action*; K. Graham, *Collectives, Classes and Revolutionary Potential in Marx*; U. Himmelstrand, *How to Become and Remain a Marxicising Sociologist. An Egocentric Report*; **Part III: On Axiology and the Socialist Project** – P. Kamolnick, *Visions of Social Justice in Marx: An Assessment of Recent Debates in Normative Philosophy*; W. Schmied-Kowarzik, *Karl Marx as a Philosopher of Human Emancipation*; H. J. Sandkühler, *Marx – Welche Rationalität? Epistemische Kontexte und Widersprüche der Transformation von Philosophie in Wissenschaft*; J. Kmita, *The*

Production of "Rational Reality" and the "Systemic Coercion"; J. Bidet, *Metastructure and Socialism*; T. Andreani, *Vers une Issue Socialiste à la Crise du Capitalisme*; W. Becker, *The Bankruptsy of Marxism. About the Historical End of a World Philosophy*; D. Aleksandrowicz, *Myth, Eschatology and Social Reality in the Light of Marxist Philosophy.*

VOLUME 61 (1997)

REPRESENTATIONS OF SCIENTIFIC REALITY
CONTEMPORARY FORMAL PHILOSOPY OF SCIENCE IN SPAIN
(Edited by Andoni Ibarra and Thomas Mormann)

Introduction – A. Ibarra, T. Mormann, *The Long and Winding Road to Philosophy of Science in Spain.* **Part 1: Representation and Measurement** – A. Ibarra, T. Mormann, *Theories as Representations*; J. Garrido Garrido, T*he Justification of Measurement*; O. Fernandez Prat, D. Quesada, *Spatial Representations and Their Physical Content*; J.A. Díez Calzada, *The Theory-Net of Interval Measurement Theory.* **Part 2: Truth, Rationality, and Method** – J.C. García-Bermejo Ochoa, *Realism and Truth Approximation in Economic Theory*; W.J.Gonzáles, *Rationality in Economics and Scientific Predictions*; J.P. Zamora Bonilla, *An Invitation to Methodonomics.* **Part 3: Logics, Semantics and Theoretical Structures** – J.L.Falguera, *A Basis for a Formal Semantics of Linguistic Formulations of Science*; A. Sobrino, E. Trillas, *Can Fuzzy Logic Help to Pose Some Problems in the Philosophy of Science?*; J. de Lorenzo, *Demonstrative Ways in Mathematical Doing*; M. Casanueva, *Genetics and Fertilization: A Good Marriage*; C.U. Moulines, *The Concept of Universe from a Metatheoretical Point of View.*

VOLUME 62 (1998)

IN THE WORLD OF SIGNS
(Edited by Jacek Juliusz Jadacki and Witold Strawiński)

Introduction. *How to Move in the World of Signs.* **Part I: Theoretical Semiotics** – A. Bogusławski, *Conditionals and Egocentric Mental Predicates*; W. Buszkowski, *On Families of Languages Generated by Categorial Grammar*; K.G. Havas, *Changing the World – Changing the Meaning. On the Meanings of the "Principle of Non-Contradiction"*; H. Hiz, *On Translation*; S. Marcus, *Imprecision, Between Variety and Uniformity: The Conjugate Pairs*; J. Peregrin, P. Sgall, *Meaning and "Propositional Attitudes"*; O.A. Wojtasiewicz, *Some Applications of Metric Space in Theoretical Linguistic.* **Part II: Methodology** – E. Agazzi, *Rationality and Certitude*; I. Bellert, *Human Reasoning and Artifical Intelligence. When Are Computers Dumb in Simulating Human Reasoning?*; T. Bigaj, *Analyticity and Existence in Mathematics*; G.B. Keene, *Taking up the Logical Slack in Natural Languages*; A. Kertész, *Interdisciplinarity and the Myth of Exactness*; J. Srzednicki, *Norm as the Basis of Form*; J.S. Stepanov, *"Cause" in the Light of Semiotics*; J.A. Wojciechowski, *The Development of Knowledge as a Moral Problem.* **Part III: History of Semiotics** – E. Albrecht, *Philosophy of Language, Logic and Semiotics*; G. Deledalle, *A Philosopher's Reply to Questions Concerning Peirce's Theory of Signs*; J. Deledalle-Rhodes, *The Transposition of Linguistic Sign in Peirce's Contributions to "The Nation"*; R. E. Innis, *From Feeling to Mind: A Note on Langer's Notion of Symbolic Projection*; R. Kevelson, *Peirce's Semiotics as Complex Inquiry: Conflicting Methods*; J. Kopania, *The Cartesian Alternative of Philosophical Thinking*; Xiankun Li, *Why Gonsung Long (Kungsun Lung) Said "White Horse Is Not Horse"*; L. Melazzo, *A Report on an Ancient Discussion*; Ding-fu Ni, *Semantic Thoughts of J. Stuart Mill and Chinese Characters*; I. Portis-Winner, *Lotman's Semiosphere: Some Comments*; J. Réthoré, *Another Close Look at the Interpretant*; E. Stankiewicz, *The Semiotic Turn of Breal's "Semantique"*; **Part IV: Linguistic** – K. Heger, *Passive and Other Voices Seen from an*

Onomasiological Point of View; L.I. Komlószi, *The Semiotic System of Events. Intrinsic Temporal and Deictic Tense Relations in Natural Language. On the Conceptualization of Temporal Schemata*; W. M. Osadnik, E. Horodecka, *Polysystem Theory, Translation Theory and Semiotics*; A. Wierzbicka, *THINK – a Universal Human Concept and a Conceptual Primitive*. **Part V: Cultural Semiotics** – G. Bettetini, *Communication as a Videogame*; W. Krysiński, *Joyce, Models, and Semiotics of Passions*; H. Książek-Konicka, *"Visual Thinking" in the Poetry of Julian Przyboś and Miron Białoszewski*; U. Niklas, *The Space of Metaphor*; M. C. Ruta, *Captivity as Event and Metaphor in Some of Cervantes' Writings*; E. Tarasti, *From Aestetics to Ethics: Semiotics Observations on the Moral Aspects of Art, Especially Music*; L. Tondl, *Is It Justified to Consider the Semiotics of Technological Artefacts?*; V. Voigt, *Poland, Finland and Hungary (A Tuatara's View)*; T.G. Winner, *Czech Poetism: A New View of Poetics Language*; J. Wrede, *Metaphorical Imagery – Ambiguity, Explicitness and Life*. **Part VI: Psycho-Socio-Semiotics** – E.M. Barth, *A Case Study in Empirical Logic and Semiotics. Fundamental Modes of Thought of Nazi Politician Vidkun Quisling, Based on Unpublished Drafts and Notebooks*; P. Bouissac, *Why Do Memmes Die?* W. Kalaga, *Threshold of Signification*, A. Podgórecki, *Do Social Sciences Evaporate?*

VOLUME 63 (1998)

IDEALIZATION IX: IDEALIZATION IN CONTEMPORARY PHYSICS
(Edited by Niall Shanks)

N. Shanks, *Introduction*, M. Bishop, *An Epistemological Role for Thought Experiments*; I. Nowak & L. Nowak, *"Models" and "Experiments" as Homogeneous Families of Notions*; S. French & J. Ladyman, *A Semantic Perspective on Idealization in Quantum Mechanics*; Ch. Liu, *Decoherence and Idealization in Quantum Measurement*; S. Hartmann, *Idealization in Quantum Field Theory*; R. F. Hendry, *Models and Approximations in Quantum Chemistry*; D. Howard, *Astride the Divided Line: Platonism, Empiricism, and Einstein's Epistemological Opportunism*; G. Gale, *Idealization in Cosmology: A Case Study*; A. Maidens, *Idealization, Heuristics and the Principle of Equivalence*; A. Rueger & D. Sharp, *Idealization and Stability:A Perspective from Nonlinear Dynamics*; D. L. Holt & R. G. Holt, *Towards a Very Old Account of Rationality in Experiment: Occult Practices in Chaotic Sonoluminescence.*

VOLUME 64 (1998)

PRAGMATIC IDEALISM. CRITICAL ESSAYS ON NICHOLAS RESCHER'S SYSTEM OF PRAGMATIC IDEALISM
(Edited by Axel Wüstehube and Michael Quante)

Introduction: A. Wüstehube, *Is Systematic Philosophy still Possible?*; T. Airaksinen, *Moral Facts and Objective Values*; L. Rodríguez Duplá, *Values and Reasons*; G. Gale, *Rescher on Evolution and the Intelligibility of Nature*; J Kekes, *The Nature of Philosophy*; P. Machamer, *Individual and Other-Person Morality: A Plea for an Emotional Response to Ethical Problems*; D. Marconi, *Opus Incertum*; M. Marsonet, *Scientific Realism and Pragmatic Idealism*; R. Martin, *Was Spinoza a Person?*; H. Pape, *Brute Facts, Real Minds and the Postulation of Reality: Resher on Idealism and the Ontological Neutrality of Experience*; J. C. Pitt, *Doing Philosophy: Rescher's Normative Methodology*; L. B. Puntel, *Is Truth "Ideal Coherence"?*; M. Quante, *Understanding Conceptual Schemes: Rescher's Quarrel with Davidson*; A. Siitonen, *The Ontology of Facts and Values*; M. Willaschek, *Skeptical Challenge and the Burden of Proof: On Rescher's Critique of Skepticism*; N. Rescher, *Responses*.

VOLUME 65 (1999)

THE TOTALITARIAN PARADIGM AFTER THE END OF COMMUNISM.
TOWARDS A THEORETICAL REASSESSMENT.
(Edited by Achim Siegel)

A. Siegel, **Introduction**: *The Changing Fortunes of the Totalitarian Paradigm in Communist Studies.* **On Recent Controversies Over The Concept Of Totalitarianism** – K. von Beyme, *The Concept of Totalitarianism – A Reassessment after the End of Communist Rule*; K. Mueller, *East European Studies, Neo-Totalitarianism and Social Science Theory*; L. Nowak, *A Conception that is Supposed to Correspond to the Totalitarian Approach to Realsocialism*; E. Nolte, *The Three Versions of the Theory of Totalitarianism and the Significance of the Historical-Genetic Version*; E. Jesse, *The Two Major Instances of Totalitarianism: Observations on the Interconnection between Soviet Communism and National Socialism.* **Classic Concept Of Totalitarianism: Reassessment And Reinterpretation** – J.P. Arnason, *Totalitarianism and Modernity: Franz Borkenau's "Totalitarian Enemy" as a Source of Sociological Theorizing on Totalitarianism*; A. Sölner, *Sigmund Neumann's "Permanent Revolution": A Forgotten Classic of Comparative Research into Modern Dictatorship*; F. Pohlmann, *The "Seeds of Destruction" in Totalitarian Systems. An Interpretation of the Unity in Hannah Arendt's Political Philosophy*; W.J. Patzelt, *Reality Construction under Totalitarianism: An Ethnomethodological Elaboration of Martin Draht's Concept of Totalitarianism*; A. Siegel, *Carl Joachim Friedrich's Concept of Totalitarianism: A Reinterpretation*; M.R. Thompson, *Neither Totalitarian nor Authoritarian: Post-Totalitarianism in Eastern Europe.*

VOLUME 66 (1999)

Leon Gumański

TO BE OR NOT TO BE? IS THAT THE QUESTION?
AND OTHER STUDIES IN ONTOLOGY, EPISTEMOLOGY AND LOGIC

Preface; *The Elements of a Judgment and Existence*; *Traditional Logic and Existential Presuppositions*; *To Be Or Not To Be? Is That The Question?*; *Some Remarks On Definitions*; *Logische und semantische Antinomien*; *A New Approach to Realistic Epistemology*; *Ausgewählte Probleme der deontischen Logik*; *An Attempt at the Definition of the Biological Concept of Homology*; *Similarity.*

VOLUME 67 (1999)

Kazimierz Twardowski

ON ACTIONS, PRODUCTS AND OTHER TOPICS IN PHILOSOPHY
(Edited by Johannes Brandl and Jan Woleński)

Introduction; Translator's Note; Self-Portrait (1926/91); Biographical Notes. **I. On Mind, Psychology, and Language:** *Psychology vs. Physiology and Philosophy (1897)*; *On the Classification of Mental Phenomena (1898)*; *The Essence of Concepts (1903/24)*; *On Idio- and Allogenetic Theories of Judgment (1907)*; *Actions and Products (1912)*; *The Humanities and Psychology (1912/76)*; *On the Logic of Adjectives (1923/27).* **II. On Truth and Knowledge:** *On So-Called Relative Truths (1900)*; A priori, *or Rational (Deductive) Sciences and* a posteriori, *or Empirical (Inductive) Sciences (1923)*; *Theory of Knowledge. A Lecture Course (1925/75).* **III. On Philosophy:** *Franz Brentano and the History of Philosophy (1895)*; *The Historical Conception of Philosophy (1912)*; *On Clear and Unclear Philosophical Style (1920)*; *Symbolomania and*

Pragmatophobia (1921); *Address at the 25th Anniversary Session of the Polish Philosophical Society (1929/31)*; *On the Dignity of the University (1933)*. **Bibliography**.

VOLUME 68 (2000)

Tadeusz Czeżowski

KNOWLEDGE, SCIENCE AND VALUES. A PROGRAM FOR SCIENTIFIC PHILOSOPHY
(Edited by Leon Gumański)

L. Gumański, *Introduction*. **Part 1: Logic, Methodology and Theory of Science** – *Some Ancient Problems in Modern Form*; *On the Humanities*; *On the Method of Analytical Description*; *On the Problem of Induction*; *On Discussion and Discussing*; *On Logical Culture*; *On Hypotheses*; *On the Classification of Sentences and Propositional Functions*; *Proof*; *On Traditional Distinctions between Definitions*; *Deictic Definitions*; *Induction and Reasoning by Analogy*; *The Classification of Reasonings and its Consequences in the Theory of Science*; *On the so-called Direct Justification and Self-evidence*; *On the Unity of Science*; *Scientific Description*. **Part 2: The World of Human Values and Norms** – *On Happiness*; *How to Understand "the Meaning of Life"?*; *How to Construct the Logic of Goods?*; *The Meaning and the Value of Life*; *Conflicts in Ethics*; *What are Values?*; *Ethics, Psychology and Logic*. **Part 3: Reality–Knowledge–World** – *Three Attitudes towards the World*; *On Two Views of the World*; *A Few Remarks on Rationalism and Empiricism*; *Identity and the Individual in Its Persistence*; *Sensory Cognition and Reality*; *Philosophy at the Crossroads*; *On Individuals and Existence*. J.J. Jadacki, *Trouble with Ontic Categories or Some Remarks on Tadeusz Czeżowski's Philosophical Views*; W. Mincer, *The Bibliography of Tadeusz Czeżowski*.

VOLUME 69 (2000)

Izabella Nowakowa, Leszek Nowak

THE RICHNESS OF IDEALIZATION

Preface; **Introduction** – *Science as a Caricature of Reality*. **Part I: THREE METHODOLOGICAL REVOLUTIONS** – *1. The First Idealizational Revolution.Galileo's-Newton's Model of Free Fall*; *2. The Second Idealizational Revolution. Darwin's Theory of Natural Selection*; *3. The Third Idealizational Revolution. Marx's Theory of Reproduction*. **Part II: THE METHOD OF IDEALIZATION** – *4. The Idealizational Approach to Science: A New Survey*; *5. On the Concept of Dialectical Correspondence*; *6. On Inner Concretization. A Certain Generalization of the Notions of Concretization and Dialectical Correspondence*; *7. Concretization in Qualitative Contexts*; *8. Law and Theory: Some Expansions*; *9. On Multiplicity of Idealization*. **Part III: EXPLANATIONS AND APPLICATIONS** – *10. The Ontology of the Idealizational Theory*; *11. Creativity in Theory-building*; *12. Discovery and Correspondence*; *13. The Problem of Induction. Toward an Idealizational Paraphrase*; *14. "Model(s) and "Experiment(s). An Analysis of Two Homogeneous Families of Notions*; *15. On Theories, Half-Theories, One-fourth-Theories, etc.*; *16. On Explanation and Its Fallacies*; *17. Testability and Fuzziness*; *18. Constructing the Notion*; *19. On Economic Modeling*; *20. Ajdukiewicz, Chomsky and the Status of the Theory of Natural Language*; *21. Historical Narration*; *22. The Rational Legislator*. **Part IV: TRUTH AND IDEALIZATION** – *23. A Notion of Truth for Idealization*; *24. "Truth is a System": An Explication*; *25. On the Concept of Adequacy of Laws*; *26. Approximation and the Two Ideas of Truth*; *27. On the Historicity of Knowledge*. **Part V: A GENERALIZATION OF IDEALIZATION** – *28. Abstracts Are Not Our Constructs. The Mental Constructs Are Abstracts*; *29. Metaphors and Deformation*; *30. Realism, Supra-Realism and Idealization*. **REFERENCES** – *I. Writings on Idealization*; *II. Other Writings*.

VOLUME 70 (2000)

QUINE. NATURALIZED EPISTEMOLOGY, PERCEPTUAL KNOWLEDGE AND ONTOLOGY
(Edited by Lieven Decock and Leon Horsten)

Introduction. **Naturalized Epistemology** – T. Derksen, *Naturalistic Epistemology, Murder and Suicide? But what about the Promises!*; Ch. Hookway, *Naturalism and Rationality*; M. Gosselin, *Quine's Hypothetical Theory of Language Learning. A Comparison of Different Conceptual Schemes and Their Logic.* **The Nature of Perceptual Knowledge** – J. van Brakel, *Quine and Innate Similarity Spaces*; D. Koppelberg, *Quine and Davidson on the Structure of Empirical Knowledge*; E. Picardi, *Empathy and Charity.* **Ontology** – S. Laugier, *Quine: Indeterminacy, 'Robust Realism', and Truth*; R. Vergauwen, *Quine and Putnam on Conceptual Relativity and Reference: Theft or Honest Toil?*; I. Douven, *Empiricist Semantics and Indeterminism of Reference*; L. Decock, *Domestic Ontology and Ideology*; P. Gochet, *Canonical Notation, Predication and Ontology.*

VOLUME 71 (2000)

LOGIC, PROBABILITY AND SCIENCE
(Edited by Niall Shanks)

N. Shanks & R.R. Gardner, *Introduction*; C. Morgan, *Canonical Models and Probabilistic Semantics (Commentary by François Lepage; Reply by Morgan)*; F. Lepage, *A Many-Valued Probabilistic Logic (Commentary by Charles Morgan; Reply by Lepage)*; P. Rawling, *The Exchange Paradox, Finite Additivity, and the Principle of Dominance (Commentary by Robert R. Gardner; Reply by Rawling)*; S. Vineberg, *The Logical Status of Conditional and its Role in Confirmation (Commentary by Piers Rawling; Reply by Vineberg)*; D. Mayo, *Science, Error Statistics, and Arguing from Error (Commentary by Susan Vineberg; Reply by Mayo)*; M.N. Lance, *The Best is the Enemy of the Good: Bayesian Epistemology as a Case Study in Unhelpful Idealization (Commentary by Leszek Nowak; Reply by Lance)*; R.B. Gardner & M.C. Wooten, *An Application of Bayes' Theorem to Population Genetics (Commentary by Lynne Seymour; Reply by Gardner and Wooten)*; P.D. Johnson, Jr., *Another Look at Group Selection (Commentary by Niall Shanks; Reply by Johnson)*; C.F. Juhl, *Teleosemantics, Kripkenstein and Paradox (Commentary by Daniel Bonevac; Reply by Juhl)*; D. Bonevac, *Constitutive and Epistemic Principles (Commentary by Mark Lance; Reply by Bonevac)*; O. Bueno, *Empiricism, Mathematical Truth and Mathematical Knowledge (Commentary by Chuang Liu; Reply by Bueno)*; Ch. Liu, *Coins and Electrons: A Unified Understanding of Probabilistic Objects (Commentary by Steven French; Reply by Liu)*; A. Maidens, *Are Electrons Vague Objects? (Commentary by David Over; Reply by Maidens).*

THE MYSTERY OF VALUES

Studies in Axiology. Ludwig Grünberg

Ed. by Cornelia Grünberg and Laura Grünberg

Amsterdam/Atlanta, GA 2000. XVIII,157 pp.
(Value Inquiry Book Series 95)
ISBN: 90-420-0670-6 Hfl. 70,-/US-$ 38.50

This study of axiology explores the axiocentricity of being human. Human beings dwell in the realm of value. Values are not simply what persons have; values in large part are what persons are. The mystique of values is analyzed here in terms of their cultural, phenomenological, and ontological status. The relationship between science and values is debated. Values should not be submitted to reductionism. Postmodernism raises new problems for the future of a philosophy of values. Yet, we may direct our hopes toward happiness, universalism, and humanism as inseparable from value-life.

HAIM GORDON:

Dwelling Poetically Educational Challenges in Heidegger's Thinking on Poetry

Amsterdam/Atlanta, GA 2000. XI,121 pp.
(Value Inquiry Book Series 94)
ISBN: 90-420-0590-4 Hfl. 55,-/US-$ 30.50

This book philosophically discusses the educational challenges of dwelling poetically, which, according to Martin Heidegger, means learning from great poems how to live a worthy life and relate authentically to beings and to Being. The gifts of great poetry are carefully described and concrete approaches are presented that the educator can adopt.

-------------------------------- ***Editions Rodopi B.V.***

USA/Canada: 6075 Roswell Rd., Ste. 219, Atlanta, GA 30328, Tel. (404) 843-4314, *Call toll-free* (U.S.only) 1-800-225-3998, Fax (404) 843-4315

All Other Countries: Tijnmuiden 7, 1046 AK Amsterdam, The Netherlands. Tel. ++ 31 (0)20 6114821, Fax ++ 31 (0)20 4472979

KATJA VERA TAVER

Johann Gottlieb Fichtes Wissenschaftslehre von 1810 Versuch einer Exegese

Amsterdam/Atlanta, GA 1999. XVIII,398 pp.
(Fichte-Studien Supplementa 12)
ISBN: 90-420-0679-X Hfl. 140,-/US-$ 77.50

Johann Gottlieb Fichtes Spätwerk - versteht man hierunter die Wissenschaftslehren zwischen 1810 und 1814 - ist weitestgehend noch unbekannt. Die von Immanuel Hermann Fichte veröffentlichte Wissenschaftslehre von 1812, auf welche die Forschung sich gern abstützt, ist vom Herausgeber teils verfälscht, die Wissenschaftslehren von 1813 und 1814 blieben unvollendet. Erst jetzt ist die Wissenschftslehre von 1810, welche Einblick gibt in Fichtes spätes Denken, zugänglich geworden.
Dieses in der J.G. Fichte Gesamtausgabe inzwischen edierte Manuskript ist jedoch nur in einer mühsamen Exegese zu entschlüsseln, da es sich um ein nacktes Gerüst für die zu haltende Vorlesung handelt. So mußte der Originaltext von 1810 im Versuch einer getreuen Exegese gegliedert und unter Beiziehung späterer Schriften und insbesondere des vorangehenden Werkes Fichtes entschlüsselt, rekonstruiert und kommentiert werden. Dies erforderte den vorsichtigen Vorgriff auf die von Immanuel Hermann Fichte edierten späten Schriften sowie den Rückgriff auf die inzwischen in kritischer Edition vorliegenden frühen Schriften Fichtes, so die "Grundlage der gesamten Wissenschaftslehre" von 1794, sowie auf die das absolute Wissen thematisierende Wissenschaftslehre von 1801/02 aus Fichtes Umbruchzeit und die Beiziehung der Wissenschaftslehre von 1804 mit der "Anweisung zum seligen Leben" von 1806, welche Fichtes Spätphase einleiten und den Boden bilden, welchem Fichtes Spätwerk entsproß. Mitberücksichtigt wurde zur Wissenschaftslehre 1810 gehörende "Wissenschaftslehre in ihrem allgemeinen Umrisse", welche Fichte selbst als Rekapitulation für seine Hörer 1810 in Druck gegeben hatte, sowie die ebenfalls zum Druck bestimmten und 1817 erschienen "Thatsachen des Bewusstseyns" von 1810/11.

------------------------------ ***Editions Rodopi B.V.***
USA/Canada: 2015 South Park Place, Atlanta, GA 30339, Tel. (770) 933-0027, *Call toll-free* (U.S.only) 1-800-225-3998, Fax (770) 933-9644
All Other Countries: Tijnmuiden 7, 1046 AK Amsterdam, The Netherlands. Tel. ++ 31 (0)20 6114821, Fax ++ 31 (0)20

JACQUES KRIEL

Matter, Mind, and Medicine
Transforming the Clinical Method

Amsterdam/Atlanta, GA 2000. XXXII,157 pp.
(Value Inquiry Book Series 93)
ISBN: 90-420-0799-0 Hfl. 70,-/US-$ 38.50

This book critically assesses the claim of modern medicine to scientific status, and it explores the implications of that claim for medicine's clinical method. Medicine models its scientific self-understanding on an obsolete positivist conception of science, reality, and consciousness. The body is viewed as a biological machine, disease as a breakdown of the machine, and therapy as the physical measures to fix that machine. Medicine's clinical method thus cannot deal with the full complexity of the ill human person. The book proposes a broader model of science and an alternative understanding of reality and of human consciousness which can become the basis of a transformed clinical method.